GLOBAL EXPOSITION OF WILDLIFE MANAGEMENT

Edited by **Gbolagade Stephen A. Lameed**

Global Exposition of Wildlife Management

http://dx.doi.org/10.5772/62974
Edited by Gbolagade Stephen A. Lameed

Contributors

Anup K.C., Shawn Larson, Fabrice Teletchea, Yasuhiko Suzuki, Jeewan Thapa, Chie Nakajima, Kamal P. Gairhe, Bhagwan Maharjan, Sarad Paudel, Susan K. Mikota, Gretchen E. Kaufman, Toshio Tsubota, Stephen V. Gordon, Deborah McCauley, Yogendra Shah, Sunday Adedoyin, Saka Jimoh, Kehinde Omifolaji

CBS, Edition **2018**

Published by InTech

Janeza Trdine 9, 51000 Rijeka, Croatia

Publishing Process Manager Martina Usljebrka

Technical Editor SPi Global

Cover Designer InTech Design Team

Additional hard copies can be obtained from orders@intechopen.com

Global Exposition of Wildlife Management, Edited by Gbolagade Stephen A. Lameed
p. cm.
Print ISBN 978-953-51-3025-3
Online ISBN 978-953-51-3026-0

Contents

Preface

This book *Global Exposition of Wildlife Management* borders on a multidisciplinary and an interdisciplinary approach to the issues on wildlife and its conservation. Wildlife is currently facing numerous threats from both natural and anthropogenic activities which are exacerbated by increasing human population. As a result, a demonstrated interdisciplinary approach from a global perspective that can guide problem solution in search of improved conservation policy and programs for species and ecosystem sustainability is necessary.

The book specifically consists of five chapters in all. Chapter 1 has two research works that deals with conservation and domestication of species from the wild written by Dr. Fabrice Teletchea. In Chapter 2 Dr Larson Shawn discusses on wildlife conservation research at accredited public aquariums in North America. Chapter 3 written by Dr Sunday Adedoyin and colleagues discusses the socioeconomic importance of wildlife with focus on bushmeat utilization in Oban Sector of Cross River National Park, Nigeria. Chapter 4 written by Mr Anup K.C. is about community forest management in Nepal. Lastly, Chapter 5 contributed by Prof Suzuki Yasuhiko and colleagues explicitly discussed on Tuberculosis within wildlife species which is emerging threat for both wildlife and humans.

Each chapter gave special reference to the prevailing problems in wildlife conservation.

The book pointed out the fact that ecotourism can be beneficial to the host community and local government, but when this is not managed in a sustainable way, it can be detrimental to the naturalness of the location. Also bush-meat utilization, which is inevitable in Africa, specifically in Nigeria, has become a major problem to biodiversity conservation. Wildlife conservation research, community forest management, integrated island management, and wildlife domestication have all been pointed out across the globe as the panacea to sustainable wildlife conservation. This book has adequately and pragmatically addressed these issues.

One cannot discuss wildlife conservation while leaving out wildlife ecology and wildlife diseases. These two aspects are very important issues of wildlife conservation. Understanding the ecology and health status of wildlife and those things that can affect their health and natural ability is a stepping stone to sustainable wildlife conservation. Hence, the book has sufficiently provided helpful information on the food and feeding strategy of vertebrates with special emphasis on food as an important factor in animal survival and also provided elaborate methods on diet studies for understanding human-wildlife conflict and conservation flagship, keystone, and umbrella species such as bears and leopard. It also highlighted tuberculosis as an emerging threat to wildlife conservation with specifics on some mammals and their endemism to South Asia.

Many species are threatened worldwide so also many unique ecosystems and improved approach to solving this problem are a panacea to successful conservation. Furthermore, to ensure continuity and sustainability of wildlife, wildlife experts cannot do it alone; hence, an interdisciplinary approach is strongly encouraged. Different experts from the field of social sciences, veterinary sciences, agricultural sciences, educational studies, and political sciences, to mention a few, put together with one vision of sustainability, and conservation can do a whole lot more. Direction from experts on different fields is needed to address this growing setback.

I appreciate the support of Publishing Process Manager Martina Usljebrka for being patient enough to encourage and direct me to put the edition right and ensure completion of the book. I also appreciate the intelligential strength of all the authors from all over the world that reported their research findings in more readable manner. Lastly, I appreciate the InTech Publisher Company for providing the opportunity to serve in this laudable capacity.

Professor Gbolagade Stephen A. Lameed
Head of Department: Wildlife and Ecotourism Management
Faculty of Agriculture and Forestry
University of Ibadan, Ibadan, Nigeria

Chapter 1

Wildlife Conservation: Is Domestication a Solution?

Fabrice Teletchea

Additional information is available at the end of the chapter

http://dx.doi.org/10.5772/65660

Abstract

Biodiversity is facing a major crisis, which is most often described as the sixth mass extinction or Anthropocene extinction. Several solutions have been proposed to save threatened animal species, among which *ex situ* conservation or captive breeding, which is the essential part of a process called domestication. The main goals of the present chapter are to define clearly what domestication is, describe what the possible consequences are and discuss whether it can truly play a significant role to save threatened animal species. Domestication appears as a possible tool to help saving threatened species. Nevertheless, the time in captive conditions has to be minimized in order to modify as less as possible wild individuals. Therefore, zoos and aquariums can play a crucial role in helping to save the most endangered species and then restore their populations in the wild, but only if they are involved in both *in situ* and *ex situ* conservation programs. More importantly, domestication should be considered as part of the solution, but not the only one, to save threatened species. The protection of wild animals *in situ*, the restoration of habitats and the development of reserves should first be considered.

Keywords: wildlife, domestication levels, endangered species, mammals, fish

1. Introduction

Of the 4 billion species estimated to have evolved on the Earth surface over the last 3.5 billion years, some 99 % are gone [1]. This illustrates how very common extinction is [1]. However, the rate of extinction of species is uneven over the course of evolution and particularly paleontologists recognize five mass extinctions as times when the Earth loses more than three-quarters of its species in a geologically short interval (typically less than 2 million years) [1]. Those big five mass extinctions are near the end of the Ordovician, Devonian, Permian, Triassic and Cretaceous periods [1]. Common features of the big five suggest that key synergies may involve unusual climate dynamics (e.g., global warming or cooling), atmospheric composition (e.g., modification of H_2S and CO_2 levels) and abnormally high-intensity ecological stressors (e.g., anoxic episodes) that affect many different lineages, among which Mammalia, Aves, Actinopterygii, Bivalvia, or Decapoda [1]. Today, it is now well accepted that biodiversity is facing a major crisis, which is most often described as the sixth mass extinction [1] or Anthropocene extinction because human impacts are at least as important as natural processes [2, 3]. One of the most obvious evidence of this biodiversity crisis is the much higher species extinction rates calculated over the past centuries than those estimated from the fossil record [1, 4]. Current extinction rates are estimated to be 1,000 higher than natural background rates of extinction (about 0.1 extinction per million species per year) and future rates are likely to be 10,000 higher [4].

Among the most charismatic endangered species for which extinction status has been formally evaluated by the International Union for Conservation of Nature (IUCN) are land and marine mammals [5]. In land, various species of equids, gomphotheres, ground sloths, glyptodonts and mammoths have already disappeared during the late Pleistocene due to humans [6, 7]. Besides, there are today numerous species that are on the brink of extinction in the wild, such as the giant panda (*Ailuropoda melanoleuca*), the black rhinoceros (*Rhinoceros bicornis*), or the tiger (*Panthera tigris*) [5]. For instance, it is estimated that there are now less than 2500 mature giant pandas in the wild, with no more than 250 mature individuals in each population [5]. The endangered and threatened marine mammals include various species of whales, manatees, dugongs, sea cows and monk seals [8–10]. Among these, only very few seem to have become globally extinct due to human activities, such as the Steller's sea cow (*Hydrodamalis gigas*) and the West Indian monk seal (*Mustela macrodon*) [10]. Numerous other animal species have also strongly declined or collapsed in the past decades due to human activities, among which many marine commercial fish [8, 9, 11]. For instance, the northern and Grand Banks Atlantic cod (*Gadus morhua*) populations have declined by more than 90 % relative to their recorded highs [11].

At global scale, the biodiversity crisis is chiefly due to overexploitation, pollution, introduction of alien species, degradation/destruction of habitats and climate change [1, 8–10, 12, 13]. Yet, hunting (or fishing) was probably the first and main reason why species became extinct in both land and marine ecosystems [7–9]. For instance, millions of large mammals, among which mammoth (*Mammuthus* sp.) or giant sloth (e.g., *Megalonyx jeffersonii*), became extinct chiefly due to hunting and then to habitat alteration [7]. In coastal ecosystems, ecological extinction caused by overexploitation (known as overfishing) clearly precedes all other pervasive anthropogenic impacts [9]. Consequently, the decline of large vertebrates as well as shellfish is first due to overfishing globally and other human impacts all come later in the standard sequence of historical events [9]. Nevertheless, habitat loss and the introduction of alien species are also causing the extinctions of many species today [13]. Indeed, habitat loss can cause species extinction when the entire habitat occupied by locally endemic species is destroyed [13]. The introduction of alien species can also cause or facilitate extinctions of native species by initiating species interactions (e.g., predation, competition) that lead to declines in the abundance and distribution of native species [13].

In the past century, two main solutions have been proposed to try to save some of the most threatened animal species, which are the creation of protected areas such as national parks [14, 15] or more recently marine reserves [16] for *in situ* conservation and *ex situ* conservation or captive breeding, which is sometimes followed by reintroduction into the wild [17, 18]. The latter solution implies the reproduction of wild animals in captivity, which is the essential part of a process called domestication. The main goals of the present chapter are thus to define clearly what domestication is, describe what the possible consequences are and discuss whether it can truly play a significant role to save threatened animal species.

2. Domestication, what does it mean?

Even though domestication is probably studied for centuries [19, 20], there is still no consensus about its definition [21–23]. Some examples [22–28] are provided in the **Table 1**. The lack of consensus on a single definition is partly due to the inherent difficulty in assigning static terms to a process involving long-term and continuous change [22]. In the present chapter, domestication is defined as a long and endless process during which captive animals become gradually adapted to both humans and captive conditions [29]. Therefore, as soon as animals are transferred from wild to captive conditions, domestication starts (**Figure 1**). If the process either voluntary or involuntary stops at this level, it corresponds to taming, i.e., to behavioral changes of a wild animal over its lifetime; yet no genetic modifications will be transmitted to the subsequent generations [28]. Once the whole life cycle is controlled in captivity (captive breeding), the process can proceed further up to the establishment of well-defined breeds displaying desired traits.

During domestication, five main genetic processes are involved in the evolution of animals [29–31]. These include two uncontrolled processes that are inbreeding and genetic drift. They result from the small size of the founder population (sometimes containing only few individuals) and create random modifications in gene frequencies. Then, the two partially controlled processes are natural selection in captivity, which results from the selection imposed on captive populations that is not due to active selection and relaxation of natural selection in captivity that can be expected to accompany the transfer from wild to captivity. The first partially controlled process, natural selection in captivity, eliminates animals incapable to reproduce in captivity and inversely favors animals that can produce a high number of offspring in the environment provided by humans [30]. In the absence of artificial selection, natural selection provides the basic selective mechanism for genetic change in captive populations [21]. The intensity of natural selection in captivity depends on the extent to which the environment allows for the development and expression of species-typical biological characteristics and on the number of generations in captivity [21]. As species possess relatively few preadaptations to captivity, natural selection is most intense during the first generations following the transition from wild to captive environments [21]. The second partially controlled process, relaxed natural selection, consists of a reduction of the selection pressure [30]. Certain behaviors important for survival in nature but not in captivity, such as food finding, predator avoidance, as well as other morphological traits (plumage or coat color), lose much of their adaptive significance in captivity [21, 30]. As a result, both genetic and phenotypic variability for these traits can thus be more variable as domestication proceeds [21, 30]. At last, the fifth genetic process is controlled, known as active selection, because changes are directional [21, 29, 30]. Artificial selection, which is the only selective mechanisms unique to domestication, involves humans selecting the breeding animals and results in the creation of different breeds [30].

Definitions	References
Domestication of wild species to produce food means that the breeding, care and feeding of organisms are more or less controlled by humans.	[24]
Domestication is defined as that process by which a population of animals becomes adapted to man and to the captive environment by some combinations of genetic changes occurring over generations and environmentally induced developmental events recurring during each generation.	[25]
Domestication involves wild animals being transformed into something more useful to humans.	[26]
The word domestication is often confusing and poorly defined, primarily because of the inherent difficulty in assigning state terms to a process involving long-term and continuous change.	[22]
The original meaning of the term domestication is the gradual adaptation of an organism to living conditions that are determined by some form of human intervention.	[27]
Domestication should not be conflated with taming. Taming is conditioned behavioral modification of an individual; domestication is permanent genetic modification of a bred lineage that leads to, among other things, a heritable predisposition toward human association.	[28]
Domestication is a continued multigenerational, mutualistic association in which one individual significantly influences the reproduction and care of another individual in order to secure a more predictable source of a resource of interest and through which the partner organism gains advantage over organisms that are not include in this relationship, thus benefiting and often improving the fitness of both the domesticator and the domesticate.	[23]

Table 1. Examples of definitions of domestication for animal species, modified after [38].

Over the course of domestication, captive animals will become domesticated (**Figure 1**). Yet, as for domestication, there is still no consensus on what a domesticated animal species is. According to most definitions [22, 32–34, 38], a domesticated species is a group of animals bred in captivity and modified from their wild ancestors (**Table 2**). However, wild/domesticated should not be considered as complementary such as true/false or dead/alive, because they represent the two extremes of a process and not a simple dichotomy [22]. In other words, no clear threshold separates wild from domesticated animals [35]. Besides, domesticated animal is neither a definitive status nor a final end point of domestication as these animals are still evolving today [30] and can sometimes return to the wild (**Figure 1**), a process known as feralization [21, 36]. According to authors, feral animals are either merely free-living individuals [36] or populations of animals (reproduce in the wild) that originated from domestic stock [21] or animals undergoing the domestication process in reverse [21]. This latter definition implies that feral animals, which are no longer exposed to artificial selection by humans or natural selection imposed by the captive environment, will therefore evolve through generations to become "wild" once more [21]. Depending on the species and the number of generations in captivity, feralization might not be possible (animals will die rapidly in nature) or will take a long period of time for animals to return to "wild" form; yet they will not go back to the original "wild" ancestor genotype and phenotype. One of the best example is cats (*Felis catus*) [28], whose domestication started about 4000 years ago from the African wildcat (*Felis silvestris lybica*) and that establishes numerous feral populations worldwide [37]. However, because feral cats are directly responsible for a large percentage of global extinctions, particularly on islands, numerous eradication programs (using trapping, hunting, poisoning and introduction of viral diseases) have been carried out in the last 30 years to preserve biodiversity, particularly seabirds [37].

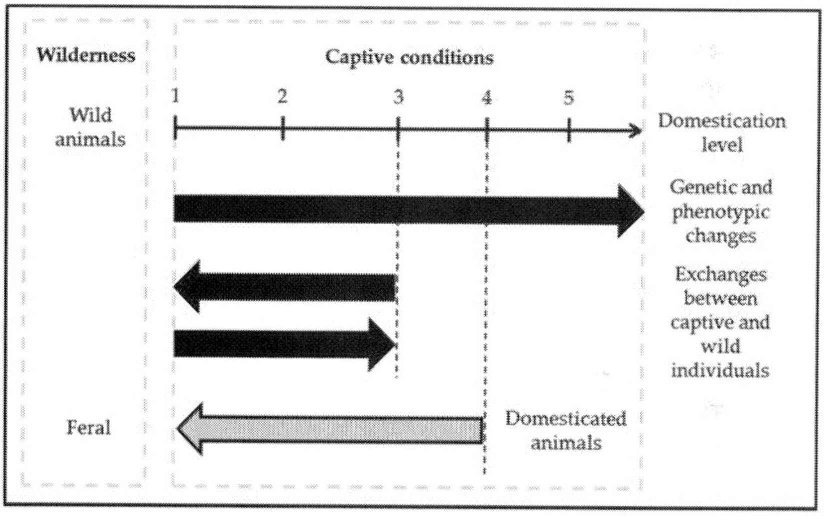

Figure 1. Evolution of a wild animal species throughout the process of domestication. As soon as wild animals are transferred to captivity (level 1), the process starts. The numbers correspond to the domestication levels described in **Table 3**. During domestication, wild animals will evolve both genetically and phenotypically, particularly when exchanges with wild congeners do no longer exist (level 4). Therefore, to minimize changes, captive animals should remain at the first three levels. Once animals have reached the level 4, they are generally considered domesticated. Domesticated animals can return to the wild and are then known as feral. The differences between feral and wild animals will depend mainly on the time spent in captivity and particularly the number of generations without exchanges with wild congeners.

In order to go beyond the usual dichotomy of wild versus domesticated animal species that was particularly not relevant for food fish production, Teletchea and Fontaine [38] created a classification based on both the level of control of the life cycle of a species in captivity and the link with wild individuals. This classification displays five levels (**Figure 1**, **Table 3**). Most authors would probably agree that at the level 4, captive animals are domesticated, particularly when they sufficiently differ from their wild ancestors [39]. Then, we applied this new concept to the fish species farmed for human consumption in order to better describe the various fish production strategies. Among the 250 species recorded in the FAO database in 2009, 70 % were classified into levels 1, 2 and 3 representing a transitory form of fish production dependent on the availability of the wild resource. In contrast, 75 species were classified at the levels 4 and 5 [38]. Yet, when a species is classified at a given level, this does not imply that the entire aquaculture production is at that level; different populations (or batches of fish) belonging to the same species can indeed display different domestication levels, even within same farm [39].

Definitions	References
A domestic animal can be defined as one that has been bred in captivity for purposes of economic profit to a human community that maintains total control over its breeding, organization of territory and food supply.	[32]
A domesticated animal species is a species bred in captivity and thereby modified from its wild ancestors in ways making it more useful to humans who control its reproduction and its food supply.	[33]
A truly domesticated species is valued and kept for a given objective, its breeding is controlled by humans, its behavior is different from its wild ancestors, its morphology and physiology display variations never observed in the wild and certain individuals at least would no longer be able to survive without human protection.	[34]
"Wild" and "domestic" represent the extremes of a process and not a simple dichotomy.	[22]
To be considered domesticated, the fish life cycle must be fully closed in captivity, independent of wild sources (domestication levels 4 and 5).	[38]

Table 2. Examples of definitions for domesticated animals, modified after [38].

Domestication level	Definitions
5	Selective breeding program is used focusing on specific goals
4	Entire life cycle closed in captivity without wild inputs
3	Entire life cycle closed in captivity with wild inputs
2	Part of the life cycle closed in captivity: several bottlenecks
1	First trials of acclimatization to the captive environment
0	Capture of wild animals (hunting or fishing)

Table 3. Domestication levels, modified after [29, 38].

In conclusion, domestication is a long and endless process during which animals become more adapted to both human and captive conditions. According to the species considered, some

http://dx.doi.org/10.5772/65660

have started this process long time ago and have thus reached the level 5 for many years or centuries, while others have just entered into it (level 1 or 2). The possible consequences are further described below for both mammals and fishes.

3. What are the main consequences of domestication?

3.1. Domestication of mammals

Domestication on land started around 12,000 years ago in at most nine areas over the world [23, 28, 33, 40]. These nine homelands of food production were Fertile Crescent, China, Mesoamerica andes/Amazonia, eastern USA, Sahel, tropical West Africa, Ethiopia and New Guinea [33]. From these primary homelands, domesticated animals were moved throughout the world, first according to an east-west axis and then a north-south axis (mainly because less evolutionary change or adaptation of domesticates was necessary for locations at the same latitudes compared to those at different latitudes) [33]. These initial introductions ultimately became the essential source of foodstuffs worldwide, resulting in that today human meat-eating diet depends on this tiny fraction of wild land mammals that were domesticated over the past millennia [26, 33, 40]. Five domesticated mammals provide the bulk of animal products (milk, meat) that are consumed across the globe. The "big five" are cow (*Bos taurus* and *B. indicus*), pig (*Sus domesticus*), sheep (*Ovis aries*), goat (*Capra hircus*) and horse (*Equus caballus*) [33].

Domestication was one of the most significant cultural and evolutionary transitions of human history [23, 28, 33, 40]. Indeed, it constitutes a core component of a major change in the way of life of an increasing number of human societies throughout the world, in a process called Neolithisation [28]. Almost everywhere in the world, hunger-gatherer communities were progressively replaced by farming societies as food production gave farmers enormous demographic, technological, political and military advantages [33]. Domestication also results in a fundamental change in the evolution of the biosphere, mainly due to the development of agriculture, which is now responsible for the transformation of approximately 40 % of the Earth's surface [41]. Today, humans are such a major geological and environmental force, at least as important as natural processes, that some considers that Earth has entered a new distinct period, called Anthropocene [3].

Over the course of evolution, wild animals were profoundly modified, including behavior, physiology, morphology and genetic [21, 23, 28, 30, 42]. One of the first modifications during domestication is behavior [21]. Yet, behavior traits did not appear or disappear, but the threshold of their expression changed [21, 30]. One of the most obvious behavioral changes manifest by all domesticates is the remarkable tolerance of proximity to (or complete lack of fear of) human [23, 28, 43]. Besides, as humans provide both protection against predators and feed, domesticated animals express a lower incidence of antipredator behaviors and show lower motivation for foraging, respectively [30]. More generally, mood, emotion, agnostic and affiliative behavior and social communication all have been modified in some way by domestication [28, 30]. Besides, most domesticated animals are more precocious than their wild counterparts [30] and the activity of their reproductive system became enhanced and relatively uncoupled from the environmental photoperiod and they all, unlike their wild ancestors, acquired the capacity to breed in any season and more often than once a year [28, 43]. At last, the most spectacular changes are probably morphological, including the overall body size (dwarfs and giants) and its proportions (fewer vertebrae, shorter tails); color, length and texture of the coat; or other manifestations of neoteny (the retention of juvenile features into sexual maturity) [28, 43]. The variation range of certain traits within a domestic species occasionally exceeds that within whole families or order, such as for dog (*Canis familiaris*) [43]. Some of these specific trait attributes (white spotting, floppy ears and curly tails) have been aptly called the morphological markers of domestication [43]. In most species, head or brain size has decreased [30]. These morphological changes may all be linked to strong selection for lowered reactivity to external stimuli [23]. More recently, the tools of molecular genetics, such as microsatellites or quantitative trait loci (QTL), have been used to investigate to what extent gene frequencies have changed between wild and domestic animals and among breeds [30, 42]. It appears that, except for certain breeds, domestic animals present a very high genetic diversity [39]. This is mainly due to the fact that exchanges between wild and captive/domestic animals were frequent in the earliest phase of domestication and probably lasted several centuries. Hence, the complete separation between wild and captive populations was relatively late and region specific [39]. It is only when breed formation started in the mid-eighteenth century, followed by the application of modern breeding methods, such as artificial insemination, in the past decades, that effective population size (N_e, which is estimated on the basis of the size of both the female and the male breeding populations) declined, resulting in strong genetic bottlenecks in certain breeds [39].

In conclusion, domestication is a very powerful process that has enabled humans to produce various domestic animals that now constitute the bulk of what we eat, i.e., cattle, pig, horse, goat and sheep. During this very long and complex process [29, 39], which started around 12,000 years ago, domesticated animals have been intensely changed resulting in numerous breeds with their own specific characteristics [42]. Besides, their numbers have increased tremendously: about 1 billion individuals for each of the big five [38]. Today, a clear dichotomy seems to exist between wild and domesticated mammalian species, which explains why researchers gave a new scientific name to some domesticated mammals [38, 42]. Nevertheless, when domesticates are sympatric with populations of the parent wild species (if the latter still exist), they can generally reproduce together [28]. Therefore, under the conceptual framework of the biological species concept, domesticated populations should not be considered as distinct species from their wild ancestors [28].

The comparison between domesticated animals and their wild ancestors is useful to study how domestication has modified animals, yet these comparisons cannot help to understand changes that happen in the first generations of domestication [30]. Only very few studies have been performed on mammalian species to evaluate early changes, among which one of the best known is on silver fox (*Vulpes vulpes*) at the Institute of Cytology and Genetics, Novosibirsk, Russia [30, 43]. During more than 50 years, about 10,500 foxes were used as parents and 50,000 offspring foxes were selected for tamability or amenability to domestication [43]. In the behavioral test, the experimenter approached the home cage, tried to open it and monitored the expression of the response [43]. The pressure of selection was very severe as less than 10 % of the tamest individuals were used as parents of the next generation [43]. As a result of such strong selection, the offspring exhibiting the aggressive and fear avoidance responses were eliminated in just two to three generations of selection [43]. After 18 generations, reproduction (capacity to breed at any time throughout the year), coat color pattern as well as other morphological traits (floppy ears, curly tails), relationships with humans and several behavioral traits (frequency of wagging, specific vocalizations, posture of the body and its communicative parts such as tails, ears and others) were very close to those of the domestic dog [30, 43]. This example demonstrates that domestication can modify very quickly captive animals [36]. Nevertheless, in this case, animals were artificially selected for tameness, which probably increased the rate of evolution of these traits in the population [30]. Besides, this study did not allow to specifically studying the transition from nature to captivity [21] as silver foxes initially used had been farm-bred for about 50 years before the experiment [43].

3.2. Fish species

Compared to land animals, the domestication of fish for human consumption has started recently [29, 38, 39, 44, 45]. Except for few species, such as the common carp (*Cyprinus carpio*) or Nile tilapia (*Oreochromis niloticus*), most trials of domesticating new fish species dated back to the early 1980s [27, 34, 39]. Consequently, there are still lots of exchanges between wild and farmed individuals and thus captive fish have only slightly changed from their wild conspecifics [46–48]. This represents a unique opportunity to study how animals evolve during the transition from wild to captive conditions, as well as during the first generations of domestication [29]. Yet, compared to the knowledge on the behavioral, physiological, morphological and genetic consequences of domestication among land vertebrates, there is very little information on the consequences of domestication in fish [49]. Besides, much knowledge acquired has been on salmonids due to their economic importance for both human consumption and sport fisheries (e.g., [50–52]).

In general, behavioral traits are among the first traits to be affected by the domestication process [46, 53]. Yet, depending on the species and captive conditions (population density, food supply, aquaria, or streams) used, it has been found that both agonistic (aggressive) and schooling behaviors could be modified (decreased or increased) during domestication [49]. While comparing wild-caught and domesticated sea bass (*Dicentrarchus labrax*) juveniles, which is one of the top farmed species in Europe [29, 45], Benhaïm et al. [53] found no differences in spatial learning. Yet, swimming behavior parameters (angular velocity, total distance traveled and velocity mean) were significantly different between the two groups. It was also found that several life-history traits have changed significantly in two salmonid species, Atlantic salmon (*Salmo salar*) and sea trout (*Salmo trutta*) reared at the hatchery at Älvkarleby, central Sweden, over the period 1968–1991 [50]. For instance, the body size of the 2-year-old smolts increased for both species and sexes. Both female and male trout adults had larger body size. Eggs were significantly larger for both species, particularly for Atlantic salmon, indicating that female salmon invest more in egg size with increasing body size. The time spent in sea has decreased for both female and male sea trout, but not for Atlantic salmon [50]. At last, several studies have been recently published that try to detect genomic differences in recently domesticated species, such as the Atlantic salmon [51]. In their study, these authors analyzed a genome-wide set of single nucleotide polymorphisms (SNPs) in three domesticated Atlantic salmon (from five to nine generations in captivity without wild inputs) and their wild conspecifics to identify loci underlying domestication. They found that the genetic differentiation between the two wild and domesticated was low and domesticated strains harbored similar level of genetic diversity compared to their wild conspecifics. Their study indicates that detecting selection in the first generations of domestication could be tricky unless selection is strong and the traits under selection show simple inheritance patterns [51]. Because a genome-wide response to selection can take many generations, as found for the Atlantic salmon, Christie et al. [52] hypothesized that the earliest changes associated with domestication may first manifest as heritable changes to global patterns of gene expression. They compared patterns of gene expression in offspring of first-generation hatchery and wild steelhead trout (*Oncorhynchus mykiss*) collected directly from the Hood River, Oregon, reared in the same environment. More than 700 genes were differentially expressed between the two groups that could not be explained by either maternal effects or by chance differences in the background levels of gene expression among unrelated families. Therefore, this is the first study to demonstrate that earliest stages of domestication are characterized by large changes in heritable patterns of gene expression, which are probably linked to adaptation to highly crowded conditions, as those genes were involved in pathways in wound healing, immunity and metabolism [52].

4. Can domestication truly help wildlife conservation?

4.1. Fisheries enhancements

Fisheries enhancements are a set of management approaches involving the use of aquaculture technologies to enhance, conserve, or restore fisheries in natural ecosystems, which are ecosystems not primarily controlled by humans, whether truly natural or modified by human activity [54–56]. Among those various aquaculture technologies, the most common form of enhancement is the release of hatchery-reared aquatic animals into natural habitats [49, 56]. Aquaculture-based enhancements have been practiced on a large scale since the mid-nineteenth century [54] and are now widely used in both inland and coastal fisheries across the world [56, 57]. For instance, state fisheries management release over 1.7 billion fish hatchery annually in the USA [56]. Besides fisheries regulation and habitat restoration, fisheries enhancements of populations are the third principal means by which fisheries can be sustained and improved [54]. Aquaculture-based enhancements can, at least in principle, increase yield through manipulation of population and/or food-web structure, aid the conservation and rebuilding of depleted or threatened populations and provide partial mitigation for ecosystem effects of fishing [54]. However, in practice, the contribution of enhancements to global fisheries has remained small [54], contrasting with the exponential growth of aquaculture in the past few decades [38]. Indeed, only a few "success stories" have been described in the literature, such as the Japanese and New Zealand scallop enhancements, Alaska salmon enhancement and Asian culture-based lake fisheries [54].

Fish cultured for fisheries enhancements enter the process of domestication as soon as they are moved from wild to captive conditions [55], which corresponds to the domestication level 1 (**Table 1**). Therefore, even though no artificial selection (selective breeding focusing on specific goals) is applied, wild fish can still be modified due to inadvertent responses to the culture environment, leading to what Lorenzen et al. [55] called "captive types." In order to mitigate as much as possible the effect of domestication and promote "wild-like types," attention should be paid to both sampling of fish for the founder population (sufficient diversity of genetic and life-history phenotypes to allow re-establishment of viable populations in the wild) and its subsequent management in captivity [55]. The most effective way of minimizing both loss of genetic diversity and the effects of domestication is to minimize the time spent in captivity [55] and release the fish at an early stage (eggs or larvae) to reduce environmental effects of the hatchery [57]. In other words, only one part of the life cycle should be controlled in captivity (level 2 in **Table 1**). Besides, the post-release performance of captive-reared fishes can be improved by modifying the captive environment of hatchery to try to mimic key aspects of natural conditions [57]. Relatively simple modifications of the captive environment, among which physical enrichment (modifications or additions of physical structure to the tanks, such as shelters) and reduced rearing density, can help produce a more wild-like fish that will perform better in the wild [57]. Yet, where populations must be maintained in captivity for multiple generations (thus reaching level 3 and perhaps level 4), there is an inherent trade-offs between the goals of maintaining diversity (avoid inbreeding and genetic drift) and minimizing adaptation because the potential for genetic adaptation is directly proportional to the heritable genetic diversity [55]. If adequate genetic diversity is maintained, it should provide sufficient reserve for feralization [55]. A recent study on Atlantic salmon demonstrated experimentally that the exposure of captive-reared fish to natural river environments during early life resulted in a twofold increase in the survivorship of offspring of wild-exposed parents compared to the offspring of captive parents [58]. The authors proposed that for lowering the possible effect of domestication, parental exposure to captivity should be minimized and exposure to the wild should be maximized but even for short period of time and within generations [58].

Salmonids are certainly the fish taxa for which most information is available on the efficiency of captive breeding programs to conserve genetic diversity and fitness of natural populations or to re-establish self-sustaining populations in the wild [59]. It appears that for most captive breeding programs, genetic diversity within populations can be sufficiently maintained in captivity for several generations. However, the captive environment may lead to unavoidable genetic changes and/or wild fitness changes in quantitative traits (despite large N_e). Decrease in fitness may potentially arise even within one generation, or after one or two generations in captivity, due to modifications, among others, of behavior, swimming performance, developmental time to hatch, embryo size, maternal reproductive investment, body morphology and age at maturity. More importantly, there is currently little empirical evidence that captive-reared lines of salmonids can be reintroduced as self-sustaining populations, particularly if the factors contributing to their initial decline are not concurrently addressed. A minimum of 15–20 years will likely be necessary to potentially achieve the conservation goal of reestablishing a self-sustaining salmonid population in the wild [59]. More recently, a comparison between first-generation hatchery-reared juvenile Atlantic salmon either released into four different river environments or kept at the hatchery showed strong differences between the two groups [60]. Hatchery fish that survived in the wild became more streamlined and more symmetrical and developed longer heads and thicker caudal peduncles and their caudal fins and opercula regenerated [60]. More importantly, this study demonstrates that hatcheries generate fish that are phenotypically mismatched to the natural environment, which may explain why they typically perform poorly in the wild [60]. Another study explored the long-term consequences of stocking captive-bred Atlantic salmon in France [61]. Based on the analysis of 11 microsatellite loci for 1428 salmon sampled from 1965 to 2006 in 25 populations, they found that the overall genetic structure among populations dramatically decreased over the period studied [61]. Depending on population, admixture rates either increased, remained stable, or decreased in samples collected between 1998 and 2006 compared to samples from 1965 to 1987, suggesting either rising, long-lasting, or short-term impacts of stocking [61]. They recommended favoring the use of native in priority as these fish may represent the most appropriate basis to restore a locally adapted population and then wild individuals (nondomesticated over several generations) to limit detrimental introgressive hybridization [61].

In conclusion, fisheries enhancement and particularly the release of captive-bred fish, might be helpful in conserving or restoring fish population [54–56, 59]. Yet, clear goals should be formulated for fish culture and domestication strategies [62], bearing in mind that different uses of fish (e.g., fish consumption versus wildlife conservation) call for very different approaches [38, 55]. At last, it should be stressed that hatchery releases should only be considered in cases where there are no realistic ways to save or maintain sensitive natural populations [57]. As a long-term strategy, habitat restoration should always be the first choice in fish conservation efforts to allow the "natural" recolonization of rivers or lakes by fish from which it has been extirpated [61].

4.2. *Ex situ* conservation: the role of zoos and aquariums

As described for fish, captive breeding of land animals is the act of bringing rare or endangered species into captivity with the hope of rearing sustained captive populations for eventual reintroduction into the wild [17]. In the past century, *ex situ* conservation programs and reintroductions of captive-bred animals have become widespread measures to protect various endangered species [18]. Successful reintroductions are exemplified by the Guam rail (*Gallirallus owstoni*), black-footed ferret (*Mustela nigripes*), California condor (*Gymnogyps californianus*) and Przewalski's horse (*Equus przewalskii*) [17, 63]. Yet, only about 10–15 % of reintroduction programs of captive-born populations were considered successful (with success judged as a self-sustaining and viable population) [17]. Although poor habitat quality is one common denominator in failed reintroductions, altered behaviors (e.g., courtship rituals, foraging/hunting routine, nest-site selection), depleted genetic diversity, or a combination of these factors also limit population growth trajectories [17, 63]. Besides, it was also found that success of translocations of wild-caught individuals that were never in captivity was much higher than those of reintroduction programs of captive-born populations, which demonstrate, as for fish species, that the best approach to minimize genetic adaptation and ensure the success of reintroduction is to reduce the time a species spends in captivity [17].

For some species reintroduction may not be an option owing to the state of their natural environment [17, 18, 64, 65]. In this case, the role of zoos and aquariums has changed from historical menageries that collect and exhibit exotic animals to modern institutions around the world that actively contribute to conservation, scientific research and public education [66–70]. Since the 1980s, many zoological gardens coordinate their breeding programs in "European Endangered Species Programs" (EEPs) and "Species Survival Plans" (SSPs) [18]. In 1993, the first World Zoo and Aquarium Conservation Strategy (WZACS) was published, which proposes clear goals for zoos and aquariums, including the need to support both *in situ* and *ex situ* conservation projects [18].

Because the goal is now to maintain a species in captivity for an extended period of time before a possible reintroduction into the wild, management strategies have to evolve [17]. Some authors proposed to attempt to minimize generations first by delaying reproduction and then by cryopreservation of germplasm [17]. Besides, because captive populations are often started with a low number of founders, either because it is difficult to collect more individuals or because there simply are no longer available, inbreeding depression is a common phenomenon in zoo populations [18]. A recent analysis showed that 67 % of *ex situ* populations in the Association of Zoos and Aquariums (AZA) institutions have a population size of less than 100 individuals and the mean N_e of Species Survival Plan (SSP) populations is 41 [18]. A perusal of studbooks from various captive mammals shows that a single founder often produces a disproportionality higher number of offspring than the remaining founders leading to a higher genetic contribution to the subsequent generations [71]. Most of the cooperatively managed breeding programs in the world possess too few captive individuals, among which too few are in proper conditions for breeding, with most often undocumented ancestries and/or too little collaboration with scientifically designated breeding recommendations [72]. These problems are leading to declining populations or decreasing gene diversity or both [72]. This demonstrates that zoo stocks still require a lot of restructuring in order to make captive breeding a more valuable contribution to species conservation [18]. In order to evaluate which methods (random mating, minimizing mean kinship and selection for docility) are the best to maintain genetic diversity in captive breeding populations, Willoughby et al. [63] evaluated genetic changes in captive populations of white-footed mice (*Peromyscus leucopus*) using microsatellites and mitochondrial DNA. After 20 generations, minimizing mean kinship resulted in slowest loss of microsatellite genetic diversity than the two others. They concluded that this method should be preferred for captive breeding, even though this method does not fully mitigate the effect of drift, as illustrated by the loss of about half of the microsatellite alleles [63]. At a much larger scale, Chargé et al. [64] found, even for a species, the Houbara bustard (*Chlamydotis undulata*), in which several thousand individuals are under a strict genetic management following worldwide used recommendations, genetic changes in several life-history traits (mean values of gamete production, body mass and courtship display rate) [64].

This example highlights the, still unresolved, question of the success of recognized guidelines for genetic management of captive populations, which in zoo are much smaller, to prevent genetic changes [64]. Other authors proposed that if there are still sufficient individuals in the wild, the immigration of individuals from the wild at a rate of one migrant per one to two generations could allow decreasing inbreeding as well as minimizing adaptation to captivity [17, 18]. Another solution that is already in practice is to exchange animals between zoos and aquariums [67, 72]. Yet, increasing regulation/restrictions on importations have reduced the ethical and logistically feasibility of importations as significant numerical support to zoo and aquarium populations [67]. At last, maintaining genetic diversity of captive populations is not sufficient as a number of difficulties in captive breeding can affect the survival of a captive population (and the success of a reintroduction program, if ever occurs). Among these known difficulties, the requirements for physical health and behavioral anomalies (e.g., stereotypic movements, lethargy, social incompatibility) have received much attention [73] and both zoos and aquariums are under increasing scrutiny for the quality of their animal management and care [67].

Today, more than 8000 species are maintained in the world's zoos and aquariums and have probably help saving the most endangered ones. Yet, in the recent decades, an increasing number of "wild" animal species are bred in captivity (tigers, gorillas and polar bears), with no longer exchanges with wild congeners [28]. Consequently, even though these species might probably not be considered as domesticated by most authors, they have reached the level 4 and therefore they could progressively diverged strongly and rapidly from their wild counterparts in few generations [36]. This could perhaps prevent possible reintroductions in the future (if habitat is restored) or at least decrease the chance of successful reintroductions. Therefore, the general objective of zoos and aquariums should not be to produce self-sufficient population, but rather to engage in the management of broader metapopulation, with carefully considered exchange between populations across a spectrum of *ex situ* to *in situ* [72]. The zoos of the future will be justified by how well they contribute positively to overall species conservation efforts, not by how well they can minimize harm to wildlife populations [72].

5. Conclusions

In 2002, Crutzen [74] coined the term Anthropocene to clearly express that since the late eighteenth century, Earth has entered a new geological epoch, dominated by human. During the past three centuries, the human population has indeed increased tenfold to more than 7 billion and the effects of humans on the global environment have escalated [74]. The most obvious environmental changes include increase of greenhouse gas concentrations, ocean acidification, alteration of global and regional nitrogen cycles, the creation of novel minerals, the transport of materials from place to place and human appropriation of net primary production [3]. During this period of time, biodiversity has been drastically modified throughout the globe due to habitat alteration/destruction, introduction of alien species and extinction of species [75]. Some even considered that truly wild nature (pristine areas) does no longer exist [14].

In this context, domestication (*ex situ* conservation) has appeared as a possible solution to save or perhaps restore populations of endangered species. Even though domestication is probably the sole solution in extreme cases, it is neither necessary nor sufficient for conserving or restoring wildlife. Indeed, numerous wild species have recovered from extreme low levels without domestication. One of best example is the rebuilding of some depleted marine fish populations by merging diverse management actions, including catch restrictions, gear modification and closed areas [76]. Obviously, the time to recovery will depend on the life-history traits of the species as well as the efficiency of conservation actions [77]. Yet, even a long-lived mammal that was heavily exploited, the humpback whale (*Megaptera novaeangliae*), which has benefited from protection from commercial whaling (since 1955), has improved from vulnerable to least concern [77]. Nevertheless, domestication has probably contributed to save some species on the brink of extinction, among which large terrestrial mammals. In the future, zoos and aquariums can play a crucial role in helping to save the most endangered species and then restore their populations in the wild, but only if they are involved in both *in situ* and *ex situ* conservation programs [72]. Otherwise, they will progressively host domesticated animals and no longer wild-type animals, which have little chance to reproduce and survive in the wild. Such a possible Noah's ark approach (focusing on a tiny proportion of wild species given the limited holding and exhibition space within and among institutions [67]) highlights that if we really want to preserve wildlife, the most important is to protect species in the wild (e.g., by reducing overfishing and poaching) as well as their environment (e.g., avoid introducing or eradicating alien species, preserve or restore "natural" habitats by establishing reserves) [14, 75, 77–79]. Stopping biodiversity decrease will thus need harmonized efforts to protect and efficiently manage critical sites, complemented by large-scale action to lower additional destruction and degradation of habitats and to encourage the sustainable use of productive ecosystems in a way that is supportive to biodiversity [77]. This implies also setting limits to human demand on nature [14]. Otherwise, natural landscapes will be progressively replaced by "working landscapes" inhabited either by domesticated or feral animals, but no longer wild animals. At the global scale, this means that we would have domesticated not only animal (and plant) species, but probably the entire Earth [15].

In conclusion, domestication appears as a powerful tool that could be useful to save threatened species. Nevertheless, the time in captive conditions has to be minimized in order to modify as less as possible wild individuals (**Figure 1**). More importantly, domestication should be considered as part of the solution, but not the only one, to save threatened species. The protection of wild animals *in situ*, the restoration of habitats and the development of reserves should first be considered.

Acknowledgements

The author would like to thank John Wiley and Sons for accepting to use the tables already published in Fish & Fisheries [38].

Author details

Fabrice Teletchea

Address all correspondence to: fabrice.teletchea@univ-lorraine.fr

University of Lorraine, UR AFPA, USC INRA, Vandoeuvre-lès-Nancy, France

References

[1] Barnosky AD, Matzke N, Tomiya S, Wogan GOU, Swartz B, Quental TB, Marshall C, McGuire JL, Lindsey EL, Maguire KC, Mersey B, Ferrer EA. Has the Earth's sixth mass extinction already arrived? Nature. 2011; 471: 51-57. DOI: 10.1038/nature09678.

[2] Wuerthner G, Crist E, Butler T, editors. Keeping the Wild. Against the Domestication of Earth. Island Press; Washington, D.C. 2014. 271 p.

[3] Corlett RT. The Anthropocene concept in ecology and conservation. Trends in Ecology and Evolution. 2015; 30: 36-41. DOI: 10.1016/j.tree.2014.10.007.

[4] De Vos JM, Joppa LN, Gittleman JL, Stephens PR, Pimm SL. Estimating the normal background rate of species extinction. Conservation Biology. 2014; 29: 452-462. DOI: 10.1111/cobi.12380.

[5] IUCN Red list. Available from http://www.iucnredlist.org/ [Accessed: 2016/07/25].

[6] Janzen DH, Martin PS. Neotropical anachronisms: the fruits the Gomphotheres ate. Science. 1982; 215: 19-27.

[7] Smith FA, Tomé CP, Elliott Smith EA, Lyons SK, Newsome SD, Stafford TW. Unraveling the consequences of the terminal Pleistocene megafauna extinction on mammal community assembly. Ecography. 2015; 38: 1-17. DOI: 10.1111/ecog.01779.

[8] Jackson JBC. What was natural in the coastal oceans? Proceedings of the National Academy of Sciences of the United States of America. 2001; 98: 5411-5418.

[9] Jackson JBC, Kirby MX, Berger WH, Bjorndal KA, Botsford LW, Bourque BJ, Bradbury RH, Cooke R, Erlandson J, Estes JA, Hughes TP, Kidwell S, Lange CB, Lenihan HS, Pandolfi JM, Peterson CH, Steneck RS, Tegner MJ, Warner RR. Historical overfishing and the recent collapse of coastal ecosystem. Science. 2001; 293: 629-638.

[10] Dulvy NK, Sadovy Y, Reynolds JD. Extinction vulnerability in marine populations. Fish and Fisheries. 2003; 4: 25-64.

[11] Hutchings JA, Baum JK. Measuring marine fish biodiversity: temporal changes in abundance, life history and demography. Philosophical Transactions of the Royal Society: B-Biological Sciences. 2005; 360: 315-338.

[12] Kerr JT, Currie DJ. Effects of human activity on global extinction risk. Conservation Biology. 1995; 9: 1258-1358.

[13] Sax DF, Gaines SD. Species diversity: from global decreases to local increases. Trends in Ecology and Evolution. 2003; 18: 561-566.

[14] Cafaro P. Expanding parks, reducing human numbers and preserving all the wild nature we can: a superior alternative to embracing the Anthropocene Era. In: Wuerthner G, Crist E, Butler T, editors. Keeping the Wild. Against the Domestication of Earth. Island Press; Washington, D.C. 2014. pp. 137-145.

[15] Wuerthner G. Why the working landscape isn't working. In: Wuerthner G, Crist E, Butler T, editors. Keeping the Wild. Against the Domestication of Earth. Island Press; Washington, D.C. 2014. pp. 162-173.

[16] Halpern BS, Warner RR. Matching marine reserve design to reserve objectives. Proceedings of the Royal Society of London B: Biological Sciences. 2003; 270: 1871-1878.

[17] Williams SE, Hoffman EA. Minimizing genetic adaptation in captive breeding programs: A review. Biological Conservation. 2009; 142: 2388-2400. DOI: 10.1016/j.biocon.2009.05.034

[18] Witzenberger KA, Hochkirch A. *Ex situ* conservation genetics: a review of molecular studies on the genetic consequences of captive breeding programmes for endangered animal species. Biodiversity Conservation. 2011; 9: 1843-1861. DOI: 10.10007/s10531-011-0074-4.

[19] Larson G, Burger J. A population genetics view of animal domestication. Trends in Genetics. 2013; 29: 197-205.

[20] Marshall FB, Dobney K, Denham T, Capriles JM. Evaluating the roles of directed breeding and gene flow in animal domestication. Proceedings of the National Academy of Sciences of the United States of America. 2014; 111: 6153-6158.

[21] Price EO. Behavioral development in animals undergoing domestication. Applied Animal Behavior Science. 1999; 65: 245-271.

[22] Dobney K, Larson G. Genetics and animal domestication: new windows on an elusive process. Journal of Zoology. 2006; 269: 261-271.

[23] Zeder MA. Core questions in domestication research. Proceedings of the National Academy of Sciences of the United States of America. 2015; 112: 3191-3198.

[24] Hale EB. Domestication and the evolution of behaviour. In: Hafez ESE, editor. The Behaviour of Domestic Animals. London: Bailliere, Tindall and Cassell; 1969. pp. 22-42.

[25] Price EO. Behavioral aspects of animal domestication. Quarterly Review Biology. 1984; 59: 1-32.

[26] Diamond J. Guns, Germs and Steel: The Fates of Human Societies. W.W. Norton & Company, New-York; 1997.

[27] Bilio M. Controlled reproduction and domestication in aquaculture – the current state of the art, Part I. Aquaculture Europe. 2007; 32: 5-14.

[28] Driscoll CA, Macdonald DW, O'Brien SJ. From wild animals to domestic pets, an evolutionary view of domestication. Proceedings of the National Academy of Sciences of the United States of America. 2009; 106: 9971-9978.

[29] Teletchea F. Domestication and genetics: what a comparison between land and aquatic species can bring? In: Pontarotti P, editor. Evolutionary Biology: Biodiversification from Genotype to Phenotype. Berlin: Springer; 2015. pp. 389-401. DOI: 10.1007/978-3-319-19932-0_20

[30] Mignon-Grasteau S, Boissy A, Bouix J, Faure J-M, Fisher AD, Hinch GN, Jensen P, Le Neindre P, Mormède P, Prunet P, Vandeputte M, Beaumont C. Genetics of adaptation and domestication in livestock. Livestock Production Science. 2005; 93: 3-14.

[31] Zeder MA. Pathways to animal domestication. In: Gepts P, Famula TR, Bettinger RL, Brush SB, Damania AB, McGuire PE, Qualset CO, editors. Biodiversity in Agriculture, Domestication, Evolution and Sustainability. Cambridge: Cambridge University Press; 2012. pp. 227-259.

[32] Clutton-Brock J. A Natural History of Domesticated Mammals. Cambridge: Cambridge University Press; 1999.

[33] Diamond J. Evolution, consequences and future of plant and animal domestication. Nature. 2002; 418: 700-707.

[34] Balon EK. About the oldest domesticates among fishes. Journal of Fish Biology. 2004; 65: 1-27.

[35] Zeder MA. Central questions in the domestication of plants and animals. Evolutionary Anthropology. 2006; 15: 105-117.

[36] Daniels TJ, Bekoff M. Feralization: the making of wild domestic animals. Behavioural Process. 1989; 19: 79-94.

[37] Nogales M, Martín A, Tershy BR, Donlan CJ, Veitch D, Puerta N, Wood B, Alonso J. A review of feral cat eradication on islands. Conservation Biology. 2004; 18: 310-319.

[38] Teletchea F, Fontaine P. Levels of domestication in fish: implications for the sustainable future of aquaculture. Fish and Fisheries. 2014; 15: 181-195.

[39] Teletchea F. Is fish domestication going too fast? Natural Resources. 2016; 7: 399-404.

[40] Mannion AM. Domestication and the origins of agriculture: an appraisal. Progress in Physical Geography. 1999. 23: 37-56.

[41] Foley JA, DeFries R, Asner GP, Barford C, Bonan G, Carpenter SR, Chapin FS, Coe MT, Daily GC, Gibbs HK, Helkowski JH, Holloway T, Howard EA, Kucharik CJ, Monfreda C, Patz JA, Prentice IC, Ramankutty N, Snyder PK. Global consequences of land use. Science. 2005; 309: 570-574.

[42] Groeneveld LF, Lenstra JA, Eding H, Toro MA, Scherf B, Pilling D, Negrini R, Finlay EK, Jianlin H, Groeneveld E, Weigend S, The GLOBALDIV Consortium. Genetic diversity in farm animals – a review. Animal Genetics. 2010; 41: 6-31.

[43] Trut L, Oskina I, Kharlamova A. Animal evolution during domestication: the domesticated fox as a model. Bioessays. 2009; 31: 349-360.

[44] Teletchea F. What is a domesticated fish? Implications for the future development of aquaculture? Ethnozootechnie. 2012; 90: 7-12.

[45] Teletchea F. Domestication of marine fish species: update and perspectives. Journal of Marine Science and Engineering. 2015; 3: 1227-1243.

[46] Olesen I, Gjedrem T, Bentsen HB, Gjerde B, Rye M. Breeding programs for sustainable aquaculture. Journal of Applied Aquaculture. 2003; 13, 179-204. DOI: 10.1300/J028v13n03_01

[47] Lind CE, Ponzoni RW, Nguyen NH, Khaw HL. Selective breeding in fish and conservation of genetic resources for aquaculture. Reproduction in Domestic Animals. 2012; 47: 255-263. DOI: 10.1111/j.1439-0531.2012.02084.x

[48] Li Y, Ponzoni RW. Some aspects of design and analysis of selection programmes in aquaculture species. Journal of Animal Breeding and Genetics. 2015; 132: 169-175.

[49] Ruzzante DE. Domestication effects on aggressive and schooling behavior in fish. Aquaculture. 1994; 120: 1-24.

[50] Pertersson E, Järvi T, Steffner NG, Ragnarsson B. The effect of domestication on some life history traits of sea trout and Atlantic salmon. Journal of Fish Biology. 1996; 48: 776-791.

[51] Mäkinen H, Vasemägi A, McGinnity P, Cross TF, Primmer CR. Population genomic analysis of early-phase Atlantic salmon (*Salmo salar*) domestication/captive breeding. Evolutionary Applications. 2015; 8: 93-107. DOI: 10.1111/eva.12230

[52] Christie MR, Marine ML, Fox SE, French RA, Blouin MS. A single generation of domestication heritably alters the expression of hundreds of genes. Nature Communications. 2016; 7: 10676. DOI: 10.1038/ncomms10676

[53] Benhaïm D, Bégout ML, Lucas G, Chatain B. Fist insight into exploration and cognition in wild caught and domesticated sea bass (Dicentrarchus labrax) in a maze. Plos One. 2013; 8: e65872. DOI: 10.1371/journal.pone.0065872

[54] Lorenzen K. Understanding and managing enhancement fisheries systems. Reviews in Fisheries Science. 2008; 16: 10-23. DOI: 10.1080/10641260701790291

[55] Lorenzen K, Beveridge MCM, Mangel M. Cultured fish: integrative biology and management of domestication and interactions with wild fish. Biological Reviews. 2012; 87: 639-700. 10.1111/j.1469-185X.2011.00215.x

[56] Lorenzen K. Understanding and managing enhancements: why fisheries scientist should care. Journal of Fish Biology. 2014; 85: 1807-1829. DOI: 10.1111/jfb.12573

[57] Johnsson JI, Brockmark S, Näslund J. Environmental effects on behavioural development consequences for fitness of captive-reared fishes in the wild. Journal of Fish Biology. 2014; 85: 1946-1971. DOI: 10.1111/jfb.12547

[58] Evans ML, Wilke NF, O'Reilly PT, Fleming IA. Transgenerational effects of parental rearing environment influence the survivorship of captive-born offspring in the wild. Conservation Letters. 2014; 7: 371-379. DOI: 10.1111/conl.12092

[59] Fraser DJ. How well can captive breeding programs conserve biodiversity? A review of salmonids. Evolutionary Applications. 2008; 1: 535-586. DOI: 10.1111/j.1752-4571.2008.00036.x

[60] Stringwell R, Lock A, Stutchbury J, Baggett E, Taylor J, Gough PJ, Garcia de Leaniz C. Maladaptation and phenotypic mismatch in hatchery-reared Atlantic salmon *Salmo salar* released in the wild. Journal of Fish Biology. 2014; 85: 1927-1945. DOI: 10.1111/jfb.12543

[61] Perrier C, Guyomard R, Baglinière JL, Nikolic N, Evanno G. Changes in the genetic structure of Atlantic salmon populations over four decades reveal substantial impacts of stocking and potential resiliency. Ecology and Evolution. 2013; 3: 2334-2349.

[62] Waters CD, Hard JJ, Brieuc MSO, Fast DE, Warheit KI, Waples RS, Knudsen CM, Bosch WJ, Naish KA. Effectiveness of managed gene flow in reducing genetic divergence associated with captive breeding. Evolutionary Applications. 2015; 8: 956-971. DOI: 10.1111/eva.12331

[63] Willoughby JR, Fernandez NB, Lamb MC, Ivy JA, Lacy RC, Dewoody A. The impacts of inbreeding, drift and selection on genetic diversity in captive breeding populations. Molecular Ecology. 2015; 24: 98-110.

[64] Chargé R, Sorci G, Saint Jalme M, Lesobre L, Hingrat Y, Lacroix F, Teplitsky C. Does recognized genetic management in supportive breeding prevent genetic changes in life-history traits? Ecological Applications. 2014; 7: 521-532. DOI: 10.1111/eva.12150

[65] Attard CRM, Möller LM, Sasaki M, Hammer MP, Bice CM, Brauer CJ, Carvalho DC, Harris JO, Beheregaray LB. A novel holistic framework for genetic-based captive-breeding and reintroduction programs. Conservation Biology. 2016; 30: 1060–1069 DOI: 10.1111/cobi.12699

[66] Kisling VN Jr, editor. Zoo and Aquarium History. Ancient Animal Collections to Zoological Gardens. Boca Raton: CRC Press; 2001. 415 p.

[67] Hutchins M, Thompson SD. Zoo and aquarium research: priority setting for the coming decades. Zoo Biology. 2008; 27: 488-497. DOI: 10.1002/zoo.20167

[68] Pelletier F, Réale D, Watters J, Boakes EH, Garant D. Value of captive populations for quantitative genetics research. Trends in Ecology and Evolution. 2009; 24: 263-270. DOI: 10.1016/j.tree2008.11.013

[69] Tlusty MF, Rhyne AL, Kaufman L, Hutchings M, McGregor Reid G andrews C, Boyle P, Hemdal J, McGilvray F, Dowd S. Opportunities for public aquariums to increase the sustainability of the aquatic animal trade. Zoo Biology. 2013; 32: 1-12. DOI: 10.1002/zoo.21019

[70] Teletchea F. Domestication level of the most popular aquarium fish species: is the aquarium trade dependent on wild populations? Cybium. 2016; 40: 21-29.

[71] Schulte-Hostedde AI, Mastromonaco GF. Integrating evolution in the management of captive zoo populations. Evolutionary Applications. 2015; 8: 413-422. DOI: 10.1111/eva.12258

[72] Lacy RC. Achieving true sustainability of zoo populations. Zoo Biology. 2013; 32: 19-26. DOI: 10.1002/zoo.21029

[73] Mason GJ. Species differences in responses to captivity: stress, welfare and the comparative method. Trends in Ecology and Evolution. 2012; 25: 713-721.

[74] Crutzen PJ. Geology of mankind. Nature. 2002; 415: 23.

[75] Martínez-Abraín A, Oro D. Preventing the development of dogmatic approaches in conservation biology: a review. Biological Conservation. 2013; 159: 539-547.

[76] Worm B, Hilborn R, Baum JK, Branch TA, Collie JS, Costello C, Fogarty MJ, Fulton EA, Hutchings JA, Jennings S, Jensen OP, Lotze HK, Mace PM, McClanahan TR, Minto C, Palumbi SR, Parma AM, Ricard D, Rosenberg AA, Watson R, Zeller D. Rebuilding global fisheries. Science. 2009; 325: 578-585. 10.1126/science.1173146

[77] Hoffmann M, et al. The impact of conservation on the status of world's vertebrates. Science. 2010; 330: 1503-1509.

[78] Kareiva P, Marvier M. What is conservation science? BioScience. 2012; 62: 962-969. DOI: 10.1525/bio.2012.62.11.5

[79] Bleich VC. Wildlife conservation and wilderness: wishful thinking? Natural Areas Journal. 2016; 36: 202-206.

Chapter 2

Wildlife Conservation Research at AZA-Accredited Public Aquariums in North America

Shawn Larson

Additional information is available at the end of the chapter

http://dx.doi.org/10.5772/66351

Abstract

Zoos and Aquariums in North America have evolved over the past few decades from institutions that exhibit animals primarily for public enjoyment to conservation organizations whose mission is to inspire and contribute significantly to wildlife conservation. The Association of Zoo and Aquariums, AZA, accredits public institutions in North America that house wildlife based on strict industry standards of exhibitry, husbandry, veterinary medicine and education. The role of conservation research has evolved in AZA facilities from just a handful of staff participating in a few projects to many facilities now having entire departments dedicated to conservation research. In 2003, AZA institutions reported in the Annual Report on Conservation Science (ARCS) that 83% of its member's participated in over 2,370 conservation projects in 107 countries and spent $77 million (AZA ARCS 2003). More recently the 2014 ARCS report stated that AZA institutions dramatically increased their conservation efforts from 2010 with 88% of members participating in conservation projects and spending 154 million. Many of the top accredited aquariums have strong conservation programs and are spending at least 3% of their budgets on conservation efforts with projects focusing on species and ecosystems and topics of global concern such as climate change and marine debris..

Keywords: conservation research, field research, American Zoo, Aquarium Association, zoos, aquariums

1. Introduction

Zoos and aquariums in the North America have evolved over the past few decades from institutions that exhibit animals for public enjoyment to conservation organizations whose mission is to inspire and contribute significantly to wildlife conservation. The Association of Zoo and Aquariums (AZA) in North America accredits public institutions that house wildlife based on industry standards of exhibitry, animal care or husbandry, veterinary medicine, and education. Founded in 1924, AZA once focused primarily on the care of captive animals and the entertainment of visitors, but now highlights the importance of field conservation work focused on saving wildlife. Although not yet an accreditation standard, AZA now encourages all member institutions to spend at least 3% of their general operating budget on field conservation directly impacting wildlife and wild ecosystems. Many accredited zoos and aquariums have general operating budgets in the tens of millions of dollars, thus even the 3% that AZA encourages institutions to spend on conservation research is substantial and equals hundreds of thousands of dollars spent per institution.

The evolution of scientific research in zoos and aquariums began slowly in the first part of the twentieth century with just a handful of institutions participating in any kind of research. Field conservation in zoos and aquariums gained momentum as society's attitudes toward wildlife and conservation changed dramatically in the later half of the century as noted by the passing of the Marine Mammal Protection Act (MMPA) in 1972 and the Endangered Species Act (ESA) in 1973. The level of conservation research in zoos and aquariums was first measured in the mid-1980s when Finlay and Maple [1] surveyed the role of research in American zoos and aquariums quoting Hediger [2] as their inspiration, "A modern zoo nowadays is not only a local place of population entertainment, but an institution which had always been indebted throughout its development to scientific inquiry, and must keep in active touch with it." By the 1980s, research in zoos and aquariums was well established with 70% of surveyed institutions stating that they participated in research or scientific activity and 59% of institutions noting it as important to their mission in official publications [1]. However, it was clear that there was room for improvement with less than half (46%) noting that their research programs were expanding, only 39% reported that they had a research committee, and only 57% published their research [1]. The largest institutions, as measured by attendance, and those affiliated with academic institutions reported the most research activities, while those reporting little to no research activity cited financial limitations and the lack of trained staff as the primary reasons for not conducting research [1].

Twelve years later the level of research in zoos and aquariums was measured again by Stoinski et al. [3]. They found that research in zoos and aquariums had increased since last surveyed in 1986. The authors noted a variety of reasons for the increase including an increased interest in conservation, the use of applied research to solve management problems, as well as an increased desire to study the wild counterparts of the animals in their collection in the field [3]. The survey found that participation in research increased to 88% of institutions with research or scientific activity noted as an objective in 83% of institutions in official publications [3]. The authors noted that since the mid-1980s the percentage of institutions conducting research increased in all facility sizes, as measured by attendance. However, they found that the percentage of institutions reporting expanding research programs decreased for all size classes except for within the largest facilities with the largest resources. In addition, the majority of zoos and aquariums surveyed in 1998 had few full time staff dedicated to research activities citing financial limitations [3]. Most notably, this survey pointed out the emergence of the importance of field research and conservation with the majority of facilities reporting studying wild counterparts of their captive collections [3].

By 2003, AZA institutions reported in the Annual Report on Conservation Science (ARCS) that 83% of its members participated in over 2370 conservation projects in 107 countries and spent $77 million [4]. Almost a decade later, the 2010 AZA ARCS report summarized that 73% of its members participated in over 1900 conservation activities and spent $130 million. In the first decade of the twenty-first century, participation as the percent of members and number of conservation projects decreased but overall funding for conservation at AZA facilities almost doubled. The reported conservation activities include donations to conservation organizations, the rescue and rehabilitation of endangered species, education programs that directly protect species in the wild, and research vital to the conservation of endangered species in the field. While this was a significant effort toward field conservation, it still reflected an investment of only 2% of AZA accredited facilities annual budgets. To increase member participation in conservation activities, in 2010 AZA launched an initiative that strongly recommended its member institutions spend at least 3% of their general operating budget on conservation research activities that directly benefit animals in the wild. This recommendation seems to have worked. The 2014 ARCS report stated that AZA institutions dramatically increased their conservation efforts from 2010 with 88% of members participating in conservation projects and spending $154 million. Although this is an improvement over the efforts reported in 2010, it is still not the approximately 200 million that would equal 3% of AZA instructional annual budgets [5].

The 2010 conservation initiative resulted in the forming of the AZA Field Conservation Committee to define and measure field conservation. This committee views AZA-accredited zoos and aquariums as conservation centers that are concerned about ecosystem health, take responsibility for species survival, contribute to research, conservation, and education, and finally provide the opportunity to develop personal connections with wild animals for society. This view is almost completely opposite that AZA had of member institutions when it first formed. The focus now is for accredited zoos and aquariums to play a more vital role than they have in the past in maintaining the planet's diverse wildlife and natural habitats while engaging the public to participate in conservation.

The AZA Field Conservation Committee defines field research as directly contributing to the long-term survival of species in natural ecosystems and habitats. Accredited institutions may meet this criteria through the following actions: direct action, defined as conducting work in the field such as research, population management, and community-based programs; species recovery, defined as food, housing, and veterinary care of animals in captive breeding programs that are slated to be reintroduced into the wild such as in rescue and rehabilitation programs; veterinary care, defined as the direct costs of wildlife disease issues for animals that can directly impact the corresponding wild population or veterinary care for wild animals not found in the collection; conservation assurance populations, defined as the direct cost of food, housing, and care of animals being housed, which cannot be reintroduced back into the wild because of poor *in situ* (in the wild) conditions, however the institution must be actively working to determine a strategy for reintroduction back into the wild; research, defined as the direct costs of research that takes place outside of the institution and helps protect species in the wild; field conservation education, defined as costs for work done on behalf of an NGO that publicizes wildlife conservation, direct costs of conservation education programs that take place outside of the institution that directly impacts the species or habitat of concern, and direct costs of training staff or volunteers not employed by the institution for field conservation work; advocacy, defined as direct costs of work done to lobby for wildlife conservation; and fundraising/grants, defined as fundraising or cash grants made to other conservation organizations to support wildlife conservation in the field.

Zoos and aquariums may impact wildlife conservation through direct actions and research. In addition to research activities, as public institutions, zoos, and aquariums play another key role in conservation. Accredited institutions in North America enjoy over 180 million visitors annually. The shift toward more conservation oriented institutions provides modern zoos and aquariums an amazing opportunity to positively impact the conservation attitudes and actions of their visitors. To measure this, in 2007, AZA conducted a nationwide study of the impacts of a visit to accredited institutions. The survey documented that after visiting an accredited zoo and aquarium in North America most adult visitors indicated a positive change in conservation attitudes and understanding [6].

The evolution of conservation in AZA accredited institutions has resulted in changing accreditation recommendations, standards, and board-approved policies. Examples of this are AZAs 2010 conservation initiative and the field conservation committee that encourage member institutions to be more conservation focused in their exhibits, educational messages, policies, and practices. For example, current AZA accreditation recommendations include that a member's mission should have a conservation theme, the institution should have a conservation program, and the institution should actively conduct conservation work. The question is how are institutions changing to meet these new standards? Specifically, how have the majority of AZA accredited aquariums in North America (here defined as the United States and Canada) responded? Most aquarium conservation research programs started very small primarily donating money to outside conservation organizations and with zero to few staff dedicated to conservation or field research. Until now, many major aquariums have their own multidisciplinary conservation programs/departments with multiple staff conducting field conservation and research.

2. Current status

The following is a summary of the current status of conservation in AZA accredited aquariums, specifically research and field conservation, in the wake of the 2010 AZA conservation initiative.

Current AZA accreditation standards suggest that all member institutions have a conservation theme. The missions of 20 AZA accredited aquariums are listed in **Table 1**. Fourteen or 70% have the word "conservation" in their mission statements. The other 30% have various words alluding to conservation such as stewardship, sustain, protect, or make a difference (**Table 1**).

Current AZA recommendations are that member institutions have a conservation program and should actively conduct field conservation work. The AZA Field Conservation Committee defines field research as directly contributing to the long-term survival of species in natural

Facility	Mission
Alaska SeaLife Center	To promote understanding and stewardship of Alaska's marine ecosystems.
Aquarium of the Bay	Explore, explain, and sustain Life.
Aquarium of the Pacific	To instill a sense of wonder, respect, and stewardship for the Pacific Ocean, its inhabitants, and ecosystems.
Audubon Aquarium of the Americas (part of the Audubon Nature Institute)	Educate our diverse audience about the natural world, enhance the care, and survival of wildlife through research and conservation.
Birch Aquarium at Scripps Institution of Oceanography	To provide ocean science education, to interpret Scripps Institution of Oceanography research, and to promote ocean conservation.
Florida Aquarium	To protect and restore our blue planet.
Georgia Aquarium	Research to contribute to the understanding of the underwater world and apply new discoveries to the conservation of aquatic life.
John G. Shedd Aquarium	To connect you to the living world, inspiring you to make a difference.
Monterey Bay Aquarium	To inspire conservation of the ocean.
Mote Marine Laboratory and Aquarium	To be a leader in nationally and internationally respected research programs that are relevant to the conservation and sustainable use of marine biodiversity, healthy habitats, and natural resources.
Mystic Aquarium	To inspire people to care for and protect our ocean planet through conservation, education, and research.
National Aquarium	Drive Marine conservation through engagement
New England Aquarium	To increase understanding of aquatic life and environments, to enable people to act to conserve the world of water, and to provide leadership for the preservation of sustainable use of aquatic resources.
Oregon Coast Aquarium	To create unique and engaging experiences that connect you to the Oregon coast and inspire ocean conservation.
Point Defiance Zoo and Aquarium	To promote and practice effective conservation on behalf of the world's wildlife.
Seattle Aquarium	Inspiring conservation of our marine environment.
SeaWorld® San Diego (research department only)	To apply basic physiological research efforts and state-of-the-art reproductive technologies toward wildlife species management and conservation.
Steinhart Aquarium	Explore, explain, and sustain life.
Tennessee Aquarium	The Tennessee Aquarium inspires wonder, appreciation and protection of water, and all life that it sustains.
Vancouver Aquarium	Conservation of aquatic life through display, communication, public programming and education, research, and direct action.

Table 1. Twenty aquarium mission statements.

Facility	F.C.	Research	Taxa and areas of focus
Alaska SeaLife Center	7	8	Climate change; marine mammals; fish; seabirds; invasive species
Aquarium of the Bay	3	1	River, bay and ocean studies; sharks
Aquarium of the Pacific	6	2	Birds; sea turtle; reefs; marine mammals; sharks; white abalone
Audubon Aquarium of the Americas (part of the Audubon Nature Institute)		2	Fish; marine mammals; sea turtles; sharks; sea horses
Birch Aquarium at Scripps Institution of Oceanography	10	NR	Marine mammals; fish; harks; sea turtle; so white abalone; coastal awareness
Florida Aquarium	7	NR	Reefs; marine mammals; sea turtle; coastal birds
Georgia Aquarium	6	NR	Penguins; sea turtles; marine mammals; sharks
John G. Shedd Aquarium	19	2	Invasive species; reptiles; marine mammals; invertebrates; fish; sea horses; penguins; sharks
Monterey Bay Aquarium	11	1	Sharks; rays; ocean conservation; fish, shorebird; seabird; sea otters; seafood watch
Mote Marine Laboratory and Aquarium	9	9	Manatees; reefs; fish; sharks; sea turtles; invertebrates; ocean acidification; toxicology
Mystic Aquarium	17	8	Amphibians; penguins; marine mammals; fish; invertebrates; reptiles
National Aquarium	15	6	Restoration; invasive species; marine animal rescue; sharks; seabirds; marine mammals
New England Aquarium	10	15	Marine mammals; sea turtles; sustainable seafood; fish; sharks; climate change; reefs; aquatic protected areas
Oregon Coast Aquarium	4	1	Invasive species; marine reserves, sea otters; marine animal rehabilitation
Point Defiance Zoo and Aquarium	4	6	Citizen science; fish; marine mammals
Seattle Aquarium	15	9	Citizen science; ocean acidification; marine mammals; reefs; fish; invertebrates; sharks; coastal monitoring
SeaWorld® San Diego (research department only)	15	NR	Marine sanctuaries; marine debris; killer whales; marine mammals; fish; sea turtles; coastal birds; seabirds
Steinhart Aquarium	8	3	Amphibians; fish; seahorses; reefs; coastal birds
Tennessee Aquarium	6	6	Fish; river ecology; amphibians; sustainable seafood
Vancouver aquarium	23	7	Amphibians; marine mammals; fish; invertebrates

Legend: F.C. = field conservation; NR = not reported.

Table 2. Number of field conservation and research projects reported in 2014 [7].

ecosystems and habitats and defines research as scientific activities that take place outside of the institution and helps protect species in the wild. All of the aquariums listed in **Table 1** reported the following conservation-related activities under field conservation (F.C.) and research in the

http://dx.doi.org/10.5772/66351

2014 ACRS report (**Table 2**). Note that the numbers of projects and species or areas of focus are likely underestimates of the breadth and scope of conservation activities as some projects may not fall into easily quantified categories.

The taxa involved in research and field conservation reported by accredited aquariums tend to reflect the facilities collections. For example, many institutions report studying coral reefs (**Figures 1** and **2**), fish and sharks (**Figures 3** and **4**), coastal birds or seabirds and marine mammals as most have all these taxa in their collections. Many aquarium exhibits, particularly those on or near the ocean, are primarily regional and thus the focus of their conservation efforts on species iconic and unique to their region (**Figure 5**). Recently, conservation projects have shifted from those with a primarily regional focus toward conservation projects with a global theme such as marine debris removal and those studying climate change.

All accredited aquariums in North America have conservation messages in their missions (**Table 1**) and all report activities that meet AZAs definition of field conservation and research (**Table 2**). Accreditation standards continue to push member institutions to do more by suggesting that all have conservation departments that are equal in breadth and scope to other major departments in aquariums such as live animal exhibits, husbandry, and education. Conservation research programs at major aquaria are organized and funded in many different ways and have evolved significantly over the years. In the past, very few organizations had conservation programs or departments. Up to now, many major accredited aquariums in North America have relatively large conservation programs/departments and many have met AZAs challenge to spend at least 3% of their general operating budget on conservation activities particularly in field conservation that directly or indirectly impacts conservation of aquatic wildlife. To determine how field research programs in AZA accredited aquariums were organized and funded, an informal survey was conducted for this paper. Forty percent of the aquariums listed in **Tables 1** and **2** responded. Survey results may be indicative of where aquariums are going regarding field conservation in that some are already where they want to be, some are in the process of getting 7there and some have just begun. As a side note even though some stated that they were happy with their institutions commitment to conservation, all agreed that more should be done for conservation of animals in the wild. The following are the results of the informal survey of major public aquariums in North America: 75% had a separately managed conservation department; 88% of those conservation departments were managed by an executive staff member at the vice president level or above; 75% of the executives managing the conservation department held a philosophical doctorate, Ph.D., and had direct experience conducting conservation science; 62% employed between 1 and 10 full time equivalent (FTE) staff, 25% surveyed employed between 10 and 20 FTE staff, and 13% employed over 20 FTE staff devoted to field conservation and/or research; 50% reported funding for their conservation departments came from grants and/or donations while the other 50% reported funding through general operating budgets; 62% reported spending more than 3% of the general operating budget on field conservation as recommended by AZA while 38% did not; finally, 50% were happy with the level of field conservation they conducted while 50% were not and felt their institutions should do more and/or have more impact on conserving wild systems.

Figure 1. Coral research: Seattle Aquarium biologists growing corals for display and to share with other aquariums to minimize taking of wild corals. Photo credit Seattle Aquarium.

Zoos and Aquariums are doing more for wildlife conservation and they may have the greatest conservation impact when they work together. In 2015 AZA launched a conservation initiative called Save Animals from Extinction or SAFE. The initiative is to encourage accredited zoos and aquariums to harness their collective resources, focus on specific endangered species, and save them from extinction by restoring healthy populations in the wild. AZA SAFE focuses on 10 signature species/groups of animals: African penguin, Asian elephant, black rhinoceros, cheetah, gorilla, whooping crane, sea turtle, sharks, vaquita, and Western pond turtle. Many of the major aquariums in **Table 2** listed conservation projects that involved either sharks and/or sea turtles as many aquariums have sharks in their collections and many either have sea turtles or participate in sea turtle rehabilitation. Another collaborative conservation initiative,

Figure 2. Aquarium researchers conducting reef surveys. Photo credit Brian McNeil.

the Aquarium Conservation Partnership (ACP) launched in 2016 as a 2-year-proof-of-concept project designed to bring the nation's leading aquariums together to achieve meaningful

Figure 3. SCUBA diver setting bait to attract free swimming sixgill sharks at the Seattle Aquarium for genetics and abundance research. Photo credit Veronica von Allworden.

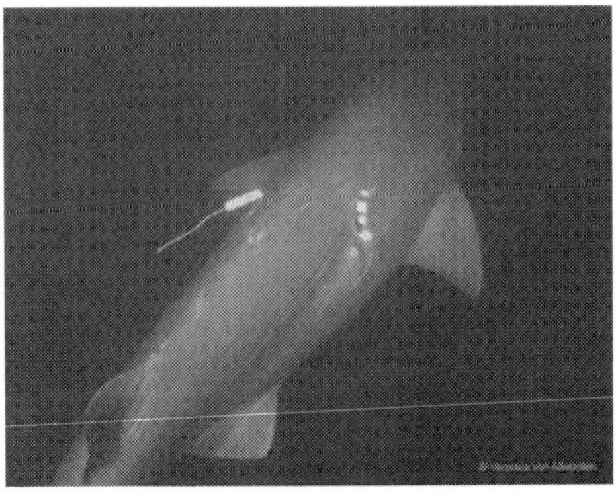

Figure 4. Free swimming sixgill shark with visual marker and acoustic tag for movement and abundance research under the Seattle Aquarium. Photo credit Veronica von Allworden.

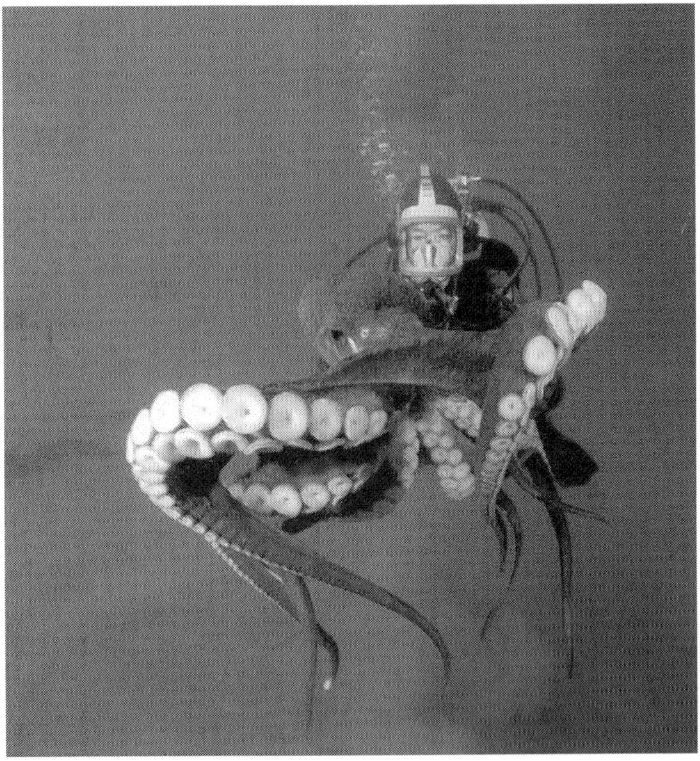

Figure 5. SCUBA diver conducting population surveys on giant Pacific octopuses. Photo credit Veronica von Allworden.

http://dx.doi.org/10.5772/66351

conservation impact for ocean and freshwater ecosystems. The ACP is a voluntary, unincorporated project that was initiated by Monterey Bay Aquarium, Shedd Aquarium, and National Aquarium. The ACP currently comprises 15 sponsoring/collaborating aquariums and ACP's first objective is to reduce ocean and freshwater plastic pollution or marine debris.

3. Conclusion

The most effective conservation research programs in public aquaria are those that use their resources wisely, partner with other institutions, and focus their efforts on field or *in situ* research that has the most impact on aquatic wildlife. Zoos and aquariums also have the opportunity to affect conservation through their visitors. The keynote speaker at AZAs 2015 annual convention was Dr. M. Sanjayan, executive vice president and senior scientist at Conservation International. In his speech he implored accredited zoos and aquariums to remember the impact they can have on conservation and to not lose sight of their conservation mission when steeped in daily activities of running a major public institution. Dr. Sanjayan calls accredited zoos and aquariums to conservation action primarily because of the unique opportunity; they have to reach the 183 million visitors a year. These institutions can directly educate and influence the kids going through their gates to understand conservation so that they grow up to make the right choices for wildlife and the environment. Virtually, every accredited zoo and aquarium has a mission focused on conservation but often the day to day running of such an operation overshadows conservation efforts. Dr. Sanjayan implored zoos and aquariums must put their mission first and foremost and focus in wildlife conservation.

Zoos and aquariums must do more for conservation and are continuing to shift focus from entertainment and educational facilities to what AZA states as "conservation centers" where real conservation work is done and animals and ecosystems are impacted in the wild. It is clear that AZA institutions are doing more than they have in the past but they must increase the breadth and scope of their field conservation programs to remain relevant in a changing world and to ensure that the amazing animals and ecosystems that they share with visitors reflect what is still in nature. Rather than being living museums to showcase what used to exist in the wild, zoos and aquarium must ensure by direct action that those species and ecosystems survive in the wild. It seems that the current trend in major North American aquariums is to move toward more direct conservation action but the fact that 50% surveyed are not happy with the level of conservation work that they are doing and almost 40% are not spending even 3% of their budgets on field conservation and research illustrates how much more needs to be done.

Author details

Shawn Larson

Address all correspondence to: s.larson@seattleaquarium.org

Seattle Aquarium, Seattle, Washington, USA

References

[1] Finlay TW, Maple TL. A survey of research in American zoos and aquariums. Zoo Biology. 1986; 5:261–268.

[2] Hediger H. Wild Animals in Captivity. New York: Dover Publications; 1964. (originally published 1950). 207 p.

[3] Stoinski TS, Lukas KE, Maple TL. A survey of research in North American zoos and aquariums. Zoo Biology. 1998; 17:167–180.

[4] The Association of Zoos and Aquariums. [Internet]. 2003. The 2003 Annual Report on Conservation and Science (ARCS). https://www.aza.org/assets/2332/2003_arcs_annual_snapshot.pdf [Accessed: 2016-07-28]

[5] The Association of Zoos and Aquariums. [Internet]. 2010. The 2010 Annual Report on Conservation and Science (ARCS). https://www.aza.org/assets/2332/2010arcs.pdf [Accessed: 2016-07-28]

[6] Falk JH, Reinhard EM, Vernon CL, Bronnenkant K, Deans NL, Heimlich JE. Why Zoos & Aquariums Matter: Assessing the Impact of a Visit. Silver Spring, MD: Association of Zoos & Aquariums; 2007.

[7] The Association of Zoos and Aquariums. [Internet]. 2014. The 2014 Annual Report on Conservation and Science (ARCS). https://www.aza.org/assets/2332/arcs_2014.pdf [Accessed: 2016-07-28]

Chapter 3

Bushmeat Utilization in Oban Sector of Cross River National Park: A 'Biodiversity Palaver'

Sunday Adedoyin, Saka Jimoh and James Omifolaji

Additional information is available at the end of the chapter

http://dx.doi.org/10.5772/66685

Abstract

This chapter examines the ecological effects of bushmeat extraction activities/methods and utilization of wildlife resources in the study area. It also highlights the perception of rural dwellers on the abundance of wildlife resources in the past 10 years as well as suggesting solution to this impending danger of depletion. Two different methods were used. The first method being stratified sampling method was used to investigate the activities involved in the extraction and utilization of wildlife resources, while the other method used a questionnaire to investigate the local people's perceptions on the abundance of wildlife resources. Results showed that gunshots were the most used (32.4%) of all the methods of extracting bushmeat in the study area, followed by the use of snares (25.5%). However, 86% (n = 86) of the respondents admitted that they consume bushmeat, while only 14% (n = 14) claimed otherwise. Buying bushmeat from markets and hunters ranked highest 55% (n = 33) among the methods of getting bushmeat in the study area, followed by the method of indirect hunting 30% (n = 18). Respondents claimed that bushmeat was occasionally consumed 38.1% (n = 37). Furthermore, 88% (n = 86) of the respondents agreed that there has been drastic change, while only about 12% (n = 12) objected to the marked difference in wildlife abundance in the park in the last 10 years. Note that 53% (n = 49) of the respondents agreed that wildlife resources in the study area have been depleted. The perception of the communities' members on the establishment of community-based wildlife management program in the area is significant (0.013*) at 0.05 level. The study revealed high level of hunting activities resulting from the use of unsustainable hunting methods, high demand for bushmeat, and lack of capacity to control hunting in the park. Lack of adequate attention to the role of bushmeat utilization as an important contributor of local livelihoods by development agencies, nongovernmental and intergovernmental organizations, and national governments contributes to the unsustainable hunting of bushmeat in tropical forests. Finally, to achieve the levels of protection necessary, habitat preservation therefore remains the key criterion for any conservation program, for without sufficient quantity and quality of habitat there will be no viable wildlife population to protect.

Keywords: biodiversity, national parks, conservation, bushmeat, hunting

1. Introduction

Forest-living people have few alternative sources of protein and income than bushmeat [1–3]. Thus, hunting of wildlife may be locally intense, and may threaten the entire populations or even species [4]. Over 160 species are endangered in West-Central Africa [5]. Uncontrolled hunting practices are only pushing these species closer to extinction. Overhunting caused the recent extinction of Miss Waldron's red colobus monkey. Several other primate species are soon to follow [6]. Up to date, it is not known exactly how many species have been eradicated and gone into extirpations due to excessive and uncontrolled hunting. However, it is evident that many species will become extinct if hunting pressures continue at their current level.

Nonetheless, the intensity of bushmeat exploitation may fluctuate even locally due to eventual perturbating events: for instance, the spreading of Ebola virus was accomplished by a considerable reduction of bushmeat consumption in West Africa [7]. High bushmeat demand has been recorded for west and central African countries (Liberia [8], Ivory Coast [9], Ghana [10, 11], Cameroon [12], and Gabon [13, 14]) and for the Congo Basin as a whole (e.g. [15–18]). Previous studies have demonstrated that bushmeat consumption is an integral part of the livelihood, both as protein requirement and important source of supplementary income, but it is also of a major sociocultural importance [19]. Overexploitation of various species has been found to be the key reason for forest wildlife declines in Africa [20]. Exploitation as a phenomenon is noted to be on the increase as a result of growing human populations, improved access to undisturbed forests, changes in hunting technology, and scarcity of alternative protein sources [21–23]. Fa et al. [24] subsequently enunciated that bushmeat depletion in the Congo Basin in real existence and supply may lead to 81% drop in less than 50 years, if the current rates of harvest continue, which may further aggravate the dramatic increase in protein malnutrition. It is therefore noteworthy to submit that overexploitation and utilization of wildlife for bushmeat and protein sources in West and Central Africa is a serious palaver which can degenerate to local, national, or worldwide extinctions of targeted species, with saddening ecological and economic consequences. Sustaining different species of wildlife both for future economic and social reasons therefore becomes a vital point of direction if the stability in the ecological system must be sustained.

The Oban hills forest area of Cross River in Nigeria is a very important conservation region with unique biodiversity, with many endangered species being confined to this region. While bushmeat trade and marketing have already been subject to research, the extent to which wildlife resources and bushmeat utilization and extraction as well as the effects on the stability of the ecosystem is yet to be substantiated. This work specifically examined the ecological effects of bushmeat extraction and utilization of wildlife resources on the stability of the ecosystem. It also highlighted the perception of rural dwellers on the abundance (increasing or dwindling) of wildlife resources for the past decades when compared to the present time as well as suggesting solution to this impending danger.

2. Location and description of the study area

Cross River National Park, the first tropical rain or moist forest national park in Nigeria, is located in Akamkpa Local Government Area of Cross River State, Nigeria. It was created by Act

http://dx.doi.org/10.5772/66685

Nos. 36 and 46 of 1991 and 46 of 1999, respectively. It covers an area of approximately 4000 km^2 and consists of two divisions: *Oban* in the south (3000 km^2) and *Okwangwo* in the north (approximately 1000 km^2). The *Oban* Division is centered on coordinates 5^{0}25′0″N 8^{0}35′0″E. CRNP is of international importance because of its unique biodiversity and species richness and endemism [25].

The study was carried out at the Oban Hill Sector. The *Oban* Division is contiguous with the Korup National Park, while the *Okwangwo* Division is contiguous with the *Takamanda* Forest Reserve, both in Cameroon. The *Oban* Hill Division of the Cross River National Park was carved out of *Oban* group Forest Reserve in 1991. It could be accessed through the Ikom-Calabar highway. The Oban sector of Cross River National Park is further divided into two corridors: the *Obong/Nsan* corridor and *Oban* corridor. Household economy in *Oban* Division is largely agrarian, although hunting, trapping, and collection of forest products are of importance for subsistence, and to an extent for trade. Economic development is seriously constrained by poor road network and market facilities. The *Oban* Hill area is inhabited predominantly by the *Ejagham* tribe with a few *Ibibio, Efiks, Calabaris,* and *Ibos*. The following are the villages where this study was carried out: *Aking/Osomba, Ifumkpa, Ekuri, Esang,* and *Ekang*. These villages have relatively large amount of tropical high forest and also consists primarily of hills and swamps. The terrain is rugged with hills ranging from 100 to more than 1000 m above sea level. Annual rainfall is estimated to range between 2500 and 3000 mm. The park is under the control of the Federal Government of Nigeria with a legal instrument promulgated through Decree No. 46 of 1991 (Act No. 46 of 1991) in the Laws of the Federal Republic of Nigeria. The flora and fauna composition of the Oban hill sector have been described by Schmidt [26], who identified 1303 species of plants, 141 lichens, and 56 mosses, 77 of which are endemic to Nigeria. Fauna biodiversity included 134 mammals, 318 birds, 42 snakes, and over 1266 butterflies. The vegetation of the Oban Sector is dominated by tropical rainforest at various stages. There are closed canopy, open canopy secondary vegetation, farm fallows, and oil palm plantations. The buffer zone consists of oil palm, cocoa, cassava, banana, plantain plantations, and maize and cocoyam farms. There are also numerous stone quarries around the buffer zone of the park.

3. Methods of data collection and analysis

Two different methods were used. The first method was used to investigate the activities involved in the extraction and utilization of wildlife resources while the other method (the use of questionnaire) was used to investigate the local people's perceptions on the abundance of wildlife resources. Oban Sector of Cross River National Park is divided into two (west and east) ranges. Five villages were purposively selected from these two ranges within the Oban Sector–Aking/Osomba (05°25'67"N, 08°38'10"E), Ekang/Esang (05°40'00"N, 08°49'00"E), Ekuri (05°21'25"N, 08°26'24"E), and Ifumkpa (05°31'56"7N, 008°17'30"4E) based on their proximity and being on the periphery of the park. Eight transects of 2 km each were laid in each of the four land-use types (core, buffer, farm fallow, and plantation) in the study locations at 1000 m intervals using a stratified sampling techniques and placed 2000 m apart to avoid animals from being detected on two neighboring transects with effective study area of 64 km^2

[27, 28]. The transects were modified as recommended by Buckland et al. [27] and were traversed for 40 consecutive times to allow for more encounters of activities of wildlife resources extraction and utilization including spent cartridges, match box, used carbide, wire snares, traps, hunting camps, hunters encountered, and gunshots heard within 10 m of both sides of the transects were recorded. Spent cartridges found were collected but wire snares were destroyed. Purposive (judgment) sampling was used as the second method to select a sample of the local respondents and study communities [29]. Purposive sampling was used due to the proximity of these villages to the park. A total number of 100 respondents from the selected villages (Aking, Ekang, Esang, Ekuri, and Ifumkpa) were interviewed, all living within a 10,000 m distance from the park boundary. The method was an interview-administered questionnaire. The questionnaire included both open-ended and fixed-response questions. The questionnaire was designed to evaluate the perceptions of local people toward wildlife resources abundance within the study area. Education and demographic information, including gender and age, were obtained from each respondent. All interviews were conducted by a research assistant who was the Assistant Range Head. Oral interviews were carried out during the day in the local language (*Ejagham and Efik*) and/or English. The average total response time was approximately 15–25 min. Stakeholders who were considered to have direct influence on the management of the park were identified and various levels of interaction were carried out. These include households, focus group discussions, village meetings, hunters, staff of the Park, members of nongovernmental organizations, and staff of Cross River State Forestry Commission and leaders of the community.

4. Data analysis

Data gathered from the questionnaire were grouped and summed by response category. The responses were recorded on a data sheet and later transcribed into English and entered into a Microsoft Excel 2010 database as well as Statistical Package for Social Sciences version 19 for Windows (IBM SPSS Inc, Chicago, USA). Where multiple responses were possible on an open-response question, data are presented as the percentage (%) of respondents giving each response, and may sum to 100%. Pearson moment correlation coefficient and descriptive statistics were also used.

5. Result presentation

Data were generated from five locations–Ifumkpa community 22 (22%), Ekuri community 19 (19%), Esang community 22 (22%), Aking/Osomba community 18 (18%), and Ekang community 19 (19%). The sample comprised of 73 males (73%) and 27 females (27%). About 32% (n = 32) of the respondents were <30 years, 54% (n = 54) were between 41 and 50 years, whereas 14% (n = 14) were older than 51 years. Forty-seven percent (n = 44) of the respondents had secondary education, 39.4% (n = 37) had primary education, 7.4% (n = 7) had NCE/OND education, 4.3% (n = 4) were illiterate, whereas 1% (n = 1) was neither educated nor illiterate.

The study revealed that 44 (47.0%) of the respondents are principally secondary school certificate holders, 37 (39.4%) are primary school leaving certificate holders, while 7 (7.4%) of the total respondents are diploma and teacher certificate holders while 1 (1.1%) of the respondents are first degree holders. Also, 4 (4.3%) of the respondents do not have formal education.

From **Table 2** it can be seen that gunshots were the most used (32.4%) of all the methods of extracting bushmeat in the study area. This was followed by the use of snares (25.5%). Other activities of extraction include hunters' trails (12.8%), hunters' camps (2.9%), and spent cartridges found (19.5%).

From **Table 3** it can be seen that 86% (n = 86) of the respondents admitted that they consume bushmeat, while only 14% (n = 14) claimed otherwise. Buying bushmeat from markets and hunters ranked highest 55% (n = 33) among the methods of getting bushmeat in the study area, followed by method of indirect hunting 30% (n = 18), which include wire snares, traps, and animal pitfalls. Other methods were through direct hunting 13.3% (n = 8) and as gift 1.7% (n = 1). The respondents claimed that bushmeat is occasionally consumed 38.1% (n = 37) in the study area, which is the highest rate of utilization/consumption. This is followed by weekly utilization/consumption 27.8% (n = 27) of bushmeat while the least utilization rate of bushmeat is fortnightly 4.1% (n = 4).

Figure 2a shows that 88% (n = 86) of the respondents agreed that there has been drastic change while only about 12% (n = 12) objected to the marked difference in wildlife abundance in the park in the last 10 years.

Figure 2b shows that 53% (n = 49) of the respondents agreed that wildlife resources in the study area has been depleted. Note that 36% (n = 33) of the respondents opined that wildlife resources are low while only 11% (10) claimed that there is still high abundance of wildlife resources in the study area.

Table 4 shows the relationship between the communities' members' perception on status of wildlife resources in the study area and the establishment of community-based wildlife management program in the area. The result shows that both variables are significant at 0.05 level (2-tailed).

6. Discussion

6.1. Demographic characteristics of respondents

Table 1 shows that young people participated more (86.0%). It also shows that majority (68.0%) of the respondents have their abode near the forest meaning that they do not have to trek much before getting into the park. This nearness then shows the peoples' dependence on the park resources. However, the result indicated that majority (73.0%) of the respondents are male (**Table 1**). **Figure 1** reveals the educational background of the respondents. Most of the respondents (47.0%) were secondary school leavers, followed by primary school leavers (39.4%). This shows that education influences their attitude and perception toward wildlife resources utilization *vis-a-vis* conservation program. This is further supported by Refs. [31, 32] that local people's knowledge about natural resources conservation are influenced by education and awareness programs, and services and benefits local people receive from conservation-related projects.

S/N	Variable	Frequency	%
1	**Location**		
	Ifumkpa	22	22.0
	Ekuri	19	19.0
	Esang	22	22.0
	Aking/Osomba	18	18.0
	Ekang	19	19.0
	Total	100	100.0
2	**Gender**		
	Male	73	73.0
	Female	27	27.0
	Total	100	100.0
3	**Age**		
	Less than 30 years	32	32.0
	41–50 years	54	54.0
	51 years and above	14	14.0
	Total	100	100.0
4	**Distance of house to the forest**		
	Near	66	68.0
	Far	31	32.0
	Total	97	100.0

Field Survey, 2013

Table 1. Demographic characteristics of respondents.

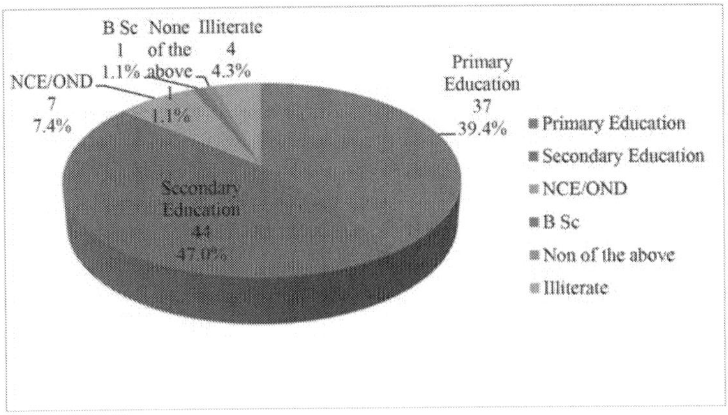

Figure 1. Educational level of respondents. Source: Field Survey, 2013.

6.2. Bushmeat extraction activities/methods in the different land-use within the study area

A total of 26 (25.5%) wire snares, 20 (19.5%) spent cartridges, 7 (6.9%) hunters encountered, 33 (32.4%) gunshots, 3 (2.9%) hunters' camps, and 13 (12.8%) were recorded for a total of 64 km of transects walked (**Table 2**). The most pronounced of all the bushmeat extraction activities was the use of guns (32.4%). The next being wire snares (19.5%). Most of the hunters set their wire snares in the rainy season. This is due to poor and very little visibility. Traps were monitored and checked between 2 and 3 days by those who set them, who were indigenes of the selected support zone communities around Cross River National Park villages. The rationale behind checking traps between 2 and 3 days was to reduce the time being invested into the traps checking, especially when quite a number of traps are set. Trap/wire snare setting is rampant during the rainy season. This is because there is less time to consummate hunting expedition. The equipment currently used by communal hunters is unsuitable for rational utilization of wildlife because it does not permit selective hunting. Communal hunters sell part of their bag at moderate prices to intermediaries who in turn supply markets, restaurants, and consumers in both rural and urban areas [33].

Bush-meat extraction activities/methods	Land-use					
	Close canopy	Secondary forest	Farm fallow	Plantation	Total	Percentage
Wire snare	9	1	15	1	26	25.5
Spent cartridges	9	4	7	0	20	19.5
Hunters encounter rate	2	0	5	0	7	6.9
Gunshots heard/guns	15	3	15	0	33	32.4
Number of hunters camp	3	0	0	0	3	2.9
Number of hunting trails	10	3	0	0	13	12.8
Total	**48**	**11**	**42**	**1**	**102**	**100.0**

Adapted from Ref. [30].

Table 2. Bushmeat extraction activities/methods in the different land-use within the study area.

6.3. The rate of utilization and consumption of bushmeat within the study area

Table 3 shows that 86% (n = 86) of the respondents admitted that they consume bushmeat, while only 14% (n = 14) claimed otherwise. Buying bushmeat from markets and hunters ranked highest 55% (n = 33) among the methods of getting bushmeat in the study area, followed by method of indirect hunting 30% (n = 18), which include wire snares, traps, and animal pitfalls. Other methods were through direct hunting 13.3% (n = 8) and as gift 1.7% (n = 1). The respondents claimed that bushmeat is occasionally consumed 38.1% (n = 37) in the study area, which is the highest rate of utilization/consumption. This is followed by weekly utilization/consumption 27.8% (n = 27) of bushmeat while the least utilization rate of bushmeat is fortnightly 4.1% (n = 4). Most household members claimed that bushmeat is significantly more delicious compared to the domesticated animals. Observations revealed that though hunting is not the major work of most communities' members, but the bushmeat from the games compliment the household nutrition sources. It is the most important source of protein for many families. However, for the household health purposes, some of the body

S/N	Variable	Frequency	%
1	**Bushmeat consumption**		
	Yes	86	86.0
	No	14	14.0
	Total	**100**	**100.0**
2	**Methods of extracting wildlife resources (bush-meat)**		
	Direct hunting	8	13.3
	Indirect hunting	18	30.0
	Buying	33	55.0
	Gift	1	1.7
	Total	**60**	**100.0**
3	**Utilization/consumption rate of bushmeat by the locals**		
	Weekly	27	27.8
	Fortnight	4	4.1
	Monthly	19	19.6
	Occasionally	37	38.1
	Never	10	10.3
	Total	**97**	**100.0**

Field Survey, 2013.

Table 3. The rate of utilization and extraction of bushmeat within the study area .

parts of the hunted animals are useful in preparation of medicinal concoctions. For instance, elephant faces and python fat are medicinal. The fat and oil of python is used as pain reliever and soothing ban, although it is also believed to be harmful and injurious to humans when swallowed. Some animal parts are also significantly important in the traditional sacrifices during healing exercise by the traditional healer. Ref. [34] recorded that bushmeat is considered as a delicacy in urban areas where people are willing to pay a premium for it, while in some areas, especially in the rural areas where they are derived, it is evident that bushmeat contributes little to the diet and more to their income [35]. Ref. [24] further asserted that the local community members perceive bushmeat as a source of long-term livelihood benefits, hence the uncontrolled and indiscriminate extraction/utilization of wildlife resources in the study study area (**Plate 1** and **Plate 2**). Other hunted wildlife species (according to personal interview with some hunters and communities' members) include the Primates (*Cercopithecus spp*), Drill (*Mandrillus leucophaeus*), *Bushbuck* (*Tragelaphus scriptus*), Pangolin (*Manis tetradactyla*), Red river hog (*Potamochoerus porcus*), and *Grasscutter* (*Thyronomys swinderianus*).

Figure 2a shows that 88% (n = 86) of the respondents agreed that there has been drastic change while only about 12% (n = 12) objected to the marked difference in wildlife abundance in the park in the last 10 years. This finding is in consonance with Jimoh et al.'s [36] submission that only two species of duiker species remain in the Oban Sector of Cross River National Park

http://dx.doi.org/10.5772/66685

Plate 1. Exhibits of *Atherurus africanus* (African brush-tailed Porcupine) and other animal species. Source: Field Survey, 2013.

Plate 2. Exhibits of *Philantomba* monticola (female blue duiker). Source: Field Survey, 2013.

which were the blue and the Ogilby's duiker. Two other large duiker species known to exist (bay and yellow-backed duikers) were not represented. Absence of these two species suggests that the species might be locally extinct, because other studies [37–39] in the area also recorded only blue and Ogilby's duikers. This is just picking duiker species for instance. The change in wildlife status and population is a menace that needs to be stemmed. Furthermore, **Figure 2b** shows that 53% (n = 49) of the respondents agreed that wildlife resources in the study area has been depleted. Note that 36% (n = 33) of the respondents opined that wildlife resources are low while only 11% (10) claimed that there is still high abundance of wildlife resources in the study area. Whichever stance the communities' members have toward wildlife population in the study area, there is depletion of wildlife resources in the study site. In consonance with the foregoing, Ref. [40] found in Ipassa Reserve, Gabon that one (bay duiker) out of the three duiker species was being locally depleted. Refs. [41, 42] in their separate studies in Udzungwa Mountains, a protected area in Tanzanian, also reported that the Abbott duiker, though previously recorded, was no longer found in the area. This is a further evidence of wildlife population depletion in Africa [43–45].

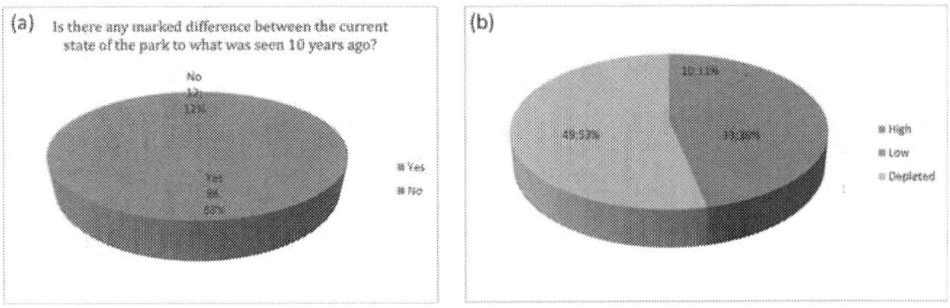

Figure 2. (a) Extent of damage done on fauna resources (bush-meat) by the local people. (b) Categories of change in wildlife resources abundance (status) in the study area. Source: Field Survey, 2013.

6.4. Perceptions of communities' members on status of wildlife and establishment of community-based wildlife management (CBWM)

Table 4 shows the relationship between the communities' members' perception on status of wildlife resources in the study area and the establishment of community-based wildlife management program in the area. The result shows that both variables are significant at 0.05 levels (2-tailed). The communities' members are aware of the dwindling population status of wildlife resources in the study area. This therefore means that for the status of wildlife to stabilize and improve, there is need to have the establishment of community-based management of wildlife resources. According to Roe [46], community wildlife management is a strategy that is based on the assumption that it is possible to improve rural livelihoods, conserve the environment, and promote economic growth. Thus, effective wildlife management models need to be developed to secure bushmeat as resource and make it available for future generations. While wildlife resources in some countries in Africa is communally owned, in most countries, wildlife is a state property and hunting is often illegal, leading to

http://dx.doi.org/10.5772/66685

a situation of low ownership and nonrecognition of user rights by the communities and even criminalization of use in the extreme cases. In order to address this situation, many countries are seeking ways to devolve and cede user rights to communities to create an incentive to invest in the long-term sustainable use of resources [47] in the form of community wildlife management arrangement.

		Perception of communities' members
Perception of communities' members	Pearson correlation	1
	N	70
Wildlife status in the study area	Wildlife status in the study area Pearson correlation	0.760
	Sig. (2-tailed)	0.021*
	N	38
Establishment of CBWM program	Establishment of CBWM program Pearson correlation	0.864
	Sig. (2-tailed)	0.013*
	N	67

*Correlation is significant at the 0.05 level (2-tailed).
Field Survey, 2013.

Table 4. Relationship between the perceptions of communities' members on status of wildlife and est.

7. Conclusion

Given the different positions on bushmeat extraction activities/methods, rate of utilization/consumption of bushmeat, indiscriminate killing of bushmeat, and perception of communities' members on establishment of community-based wildlife management, the study indicated high level of hunting activities resulting from the use of unsustainable hunting methods, high demand for bushmeat, and lack of capacity to control hunting in the park. Large mammals are especially exposed and become easy targets for hunters. Changes in traditional hunting practices through the use of improved hunting technology have decreased the likelihood that hunting will be sustainable. Therefore, protecting wildlife through protected area systems and development of biodiversity and corridors is the method which has the greatest potential to address global concerns regarding wildlife populations in the Oban Sector of Cross River National Park and West Africa at large. Alternatively, sport hunting and ecological tourism could serve as potent weapon in arresting the downward slide of wildlife resources in the park. To be truly effective, these kinds of efforts would have to be done in full collaboration with local communities, addressing issues of access and user rights, enforcement concerns, and lack of adequate data on population dynamics. Finally, to achieve the levels of protection necessary, habitat preservation therefore remains the key criterion for any conservation program, for without sufficient quantity and quality of habitat there will be no viable wildlife population to protect.

Acknowledgements

We are grateful to the Volkswagen Foundation, Hanover, Germany, for providing the grant to the University of Ibadan, Nigeria, within its 'Africa Initiative'- 'Knowledge for Tomorrow Cooperative Research Projects in Sub-Saharan Africa' which made this study possible. We thank Federal Board for National Park Service, Abuja, the Conservator of Cross River National Park and other members of staff of the Cross River National Park, for their kind assistance in planning and logistics.

Author details

Sunday Adedoyin[1]*, Saka Jimoh[2] and James Omifolaji[3]

*Address all correspondence to: adedoyin.so.mnim@gmail.com

1 University of Benin, Benin, Nigeria

2 University of Ibadan, Ibadan, Nigeria

3 Federal University, Dutse, Nigeria

References

[1] Bakarr, M. I., O. Ampadu-Agyei, E. Adomako & R. Ham. (2002) Bushmeat utilisation, human livelihoods and conservation of large mammals in West Africa. pp. 45–53. *In*: S. A. Mainka & M. Trivedi (eds.) *Links between Biodiversity Conservation, Livelihoods and Food Security: The Sustainable Use of Wild Species for Meat.* Gland, IUCN.

[2] Mainka, S. A. & M. Trivedi. (2002) *Links between Biodiversity Conservation, Livelihoods and Food Security: the Sustainable Use of Wild Species for Meat.* IUCN, Gland.

[3] Nasi, R., D. Brown, D. Wilkie, E. Bennett, C. Tutin, G. van Tol & T. Christophersen. (2008) *Conservation and Use of Wildlife-Based Resources: the Bushmeat Crisis.* Technical Series No. 33, Secretariat of the Convention on Biological Diversity, Montreal, and Center for International Forestry Research (CIFOR), Bogor.

[4] Oates, J. F. (2011) *Primates of West Africa.* Conservation International, New York.

[5] Hoffman, M. & J. Pilgrim .(2005) *All Threatened Vertebrate Species Occurring in Guinean Forests of West Africa.* Retrieved April 3, 2005, from Conservation International, SpeciesDatabaseWebpages:http://www.biodiversityhotspots.org/xp/Hotspots/west_africa/

[6] Oates, J. F., M. Abedi-Lartey, W.S. McGraw, T. T. Struhsaker & G. Whitesides. (2000) Extinction of a West African red colubus monkey. *Conservation Biology* 14(5), 1526–1532.

[7] Akani, G. C., D. Dendi & L Luiselli. (2015) Ebola virus effects on the bush-meat trade in West Africa. *African Journal of Ecology*, doi: 10.1111/aje.12231

[8] Ajayi, S. S. (1979) *Utilisation of Forest Wildlife in West Africa*. Misc/79/26, Rome, FAO.

[9] Caspary, H. U. (2001) Regional dynamics of hunting and bush-meat utilization in West Africa – An overview. pp. 11–16. *In*: M. I. Bakarr, G. A. B.de Fonseca, R. A. Mittermeier, A. B. Rylands & K. W. Painemilla (eds.) *Hunting and Bushmeat Utilization in the African Rain Forest. Perspectives Toward a Blueprint for Conservation Action*. Conservation International, Washington DC.

[10] Cowlishaw, G., S. Mendelson & J. M. Rowcliffe. (2007) Livelihoods and Sustainability in a Bush-meat Commodity Chain in Ghana. pp. 32–46. *In*: G. Davies & D. Brown (eds.) *Bush-meat and Livelihoods: Wildlife Management and Poverty Reduction*. Blackwell Publishing, Oxford.

[11] de Vos, A. (1978) Game as food: A report on its significance in Africa and Latin America. *Unasylva* 29: 2–12.

[12] Muchaal, P. K. & G. Ngandjui. (1999) Impact of village hunting on wildlife populations in the Western Dja Reserve, Cameroon. *Conservation Biology* 13: 385–396.

[13] Lahm, S. A. (1996) Gabon's village hunting: Assessing its impact. *African Primates* 2: 23–24.

[14] Steel, E. A. (1994) *Study of the Value and Volume of Bush-meat Commerce in Gabon*. WWF Programme pour le Gabon. WWF, Libreville, Gabon.

[15] Draulans, D. & E. Van Krunkelsven. (2002) The impact of war on forest areas in the Democratic Republic of Congo. *Oryx* 36: 35–40.

[16] Vanwijnsberghe, S. (1996) *Etude Sur la Chasse Villageoise Aux Environs au Parc d' Odzala*. Wildlife Management Working Paper Number 5, FAO, Rome.

[17] Wilkie, D. S., & J. F. Carpenter. (1999) Bush-meat hunting in the Congo Basin: An assessment of impacts and options for Mitigation. *Biodiversity and Conservation* 8: 927–955.

[18] Wilkie, D. S., M. Starkey, K. Abernethy, E. Effa Nsame, P. Telfer & R. Godoy. (2005) Role of species and wealth consumer demand for bush-meat in Gabon, Central Africa. *Conservation Biology* 19: 268–274.

[19] de Merode, E., K .Homewood & G. Cowlishaw. (2003) *Wild Resources and Livelihoods of Poor Households in the Democratic Republic of Congo*. ODI, Wildlife Policy Briefing, No 1, London, UCL.

[20] Bennett, E.L., J.G. Robinson, E.J. Milner-Gulland, M. Bakarr, ,H.E. Eves & D.S. Wilkie. (2007) Hunting the World's Wildlife to Extinction. *Oryx*, 36, 328–329. http://dx.doi.org/10.1017/S0030605302000637

[21] Robinson, J.G. & E.L. Bennett. (1999) *Hunting for Sustainability in Tropical Forest*. Columbia University Press, New York.

[22] Robinson, J.G. & E.L. Bennett. (2000) Carrying capacity limits to sustainable hunting in tropical forests. *In:* J.G. Robinson & E.L. Bennett (eds.) *Hunting for Sustainability in Tropical Forests*. Columbia University Press, New York.

[23] Fa, J.E., C.A. Peres & J.A. Meeuwig. (2002) Bush-meat exploitation in tropical forests: An intercontinental comparison. *Conservation Biology* 16: 232–237.

[24] Fa, J.E., C.A. Peres & J.A. Meeuwig. (2003) Bush-meat and food security in the Congo-Basin: Linkages between wildlife and people's future. *Environmental Conservation* 30: 71–78.

[25] Myre, N., R. Mittermeier, C. Mittermeier, G. Fonseca, & J. Kent. (2000) Biodiversity hotspots for conservation priorities. *Nature* 403(6772): 853–858.

[26] Schmidt, K. (1996) *Botanical Survey in the Oban Division of CRNP*. Technical Report on Oban Hill program, Calabar. pp 1–55.

[27] Buckland, S.T., D.R. Anderson K.P. Burnham, J.L. Laake, D.L. Borchers, & L. Thomas. (2001) *Introduction to Distance Sampling: Estimating Abundance of Biological Populations*. Oxford University Press, Oxford.

[28] Waltert, M., S. Heber, S. Riedelbauch, J.L. Lien & M .Muhlenberg. (2006) Estimates of Blue Duiker (*Cephalophus monticola*) densities from diurnal and nocturnal line transects in the Korup Region, South-Western Cameroon. *African Journal of Ecology*, 44, 290–292. http://dx.doi.org/10.1111/j.1365-2028.2006.00631.x

[29] Tongco, M.D.C. (2007) Purposive sampling as a tool for informant selection. *Ethnobotany Research & Applications*, 5, 147–158.

[30] Lameed, G.A., J.K. Omifolaji, A.S. Abere & S.O. Ilori. (2015) Hunting intensity on wildlife population in Oban sector of Cross River National Park. *Natural Resources*, 6, 325–330. http://dx.doi.org/10.4236/nr.2015.64029

[31] Jalilova, G. & H. Vacik. (2012) Local people's perceptions of forest biodiversity in the Walnut Fruit Forests of Kyrgyzstan. *International Journal of Biodiversity Science, Ecosystem Services & Management*, 8, 204–216. http://dx.doi.org/10.1080/21513732.2012.696557

[32] Newmark, W.D., N.L. Leonard, H.I. Sariko, & D.G.M. Gamassa. (1993) Conservation attitudes of local people living adjacent to five protected areas in Tanzania. *Biological Conservation*, 63, 177–183. http://dx.doi.org/10.1016/0006–3207(93)90507-W

[33] Caspary, H.U. 1999. Wildlife Utilization in Côte d'Ivoire and West Africa – Potentials and Constraints for Development Cooperation. Tropical Ecology Support Program, Eschborn, Germany.

[34] Bowen-Jones, E., D. Brown & E. J. Robinson. (2002) *Assessment of the Solution Orientated Research Needed to Promote a More Sustainable Bushmeat Trade in Central and West Africa*. Report to the Wildlife and Countryside Directorate, DEFRA, London.

[35] de Merode, E., K. Homewood & G. Cowlishaw. (2004) The value of Bush-meat and other wild foods to rural households living in extreme poverty in Democratic Republic of Congo. *Biological Conservation,* 118: 573–581.

[36] Jimoh, S.O., E.T. Ikyaagba, A.A. Alarape, A. A. Adeyemi & M. Waltert. (2013) Local depletion of two larger Duikers in the Oban Hills Region, Nigeria. *African Journal of Ecology.*

[37] Oates, J.F., R.A. Bergl, & J.M. Linder. (2004) *Africa's Gulf of Guinea Forests: Biodiversity Patterns and Conservation Priorities: Advances in Applied Biodiversity Science,* number 6. Conservation International, Washington DC. 2–34 pp.

[38] Fa, J.E., S. Seymour, J. Dupain, R. Amin, L. Albrechtsen & D. Mcdonald. (2006) Getting to grips with the Magnitude of Exploitation: Bush-meat in the Cross-Sanaga Rivers Region, Nigeria and Cameroun. *Biological Conservation* 129: 497–510.

[39] Eniang, E.A., M.E. Eniang & C.E. Akpan. (2008) Bush-meat trading in the Oban Hills Region of South-Eastern Nigeria: Implications for sustainable livelihoods and conservation. *Ethiopian Journal of Environmental Studies and Management* 1: 70–83.

[40] Van Vliet, N., R. Nasi, L. Emmons, F. Feer, P. Mbazza & M. Bourgarel. (2007) Evidence for the local depletion of bay duiker *Cephalophus dorsalis,* within the Ipassa man and Biosphere Reserve, North-East Gabon. *African Journal of Ecology* 45: 440–443.

[41] Rovero, F. & A.R. Marshall. (2004) Estimating the Abundance of Forest Antelopes by line-transect techniques: A case from the Udzungwa Mountains, Tanzanian. *Tropical Zoology* 17: 267–277.

[42] Niesen, M.R. (2006) Importance, cause and effect of bush-meat hunting in the Udzungwa Mountains, Tanzanian: Implications for community-based wildlife management: *Biological Conservation* 128: 509–516.

[43] Brashares, J.S., P. Areese, & M.K. Sam. (2001) Human demography and reserve size predict wildlife extinction in West Africa. *Proceedings of the Royal Society of London* B 268: 2473–2478.

[44] Western, D., S. Russell & I. Cuthill. (2009) The status of wildlife in protected areas compared to non-protected areas of Kenya. *PLoS ONE* 4(7): 6140. doi: 10.137/journal.pone.0006140

[45] Craigie, I.D., J.E.M. Baillie, A. Balmford, C. Carbone, B. Collen, R.E. Green & J.M. Hutton. (2010) Large mammal population declines in Africa's protected areas. *Biological Conservation* 143: 2221–2228.

[46] Roe, D. (2001) *Community-based wildlife management: Improved livelihoods and wildlife conservation.* Bio-Brief, No 1. International Institute for Environment and Development, London.

[47] Pailler, S. (2005) The necessity, complexity and difficulty of resolving the bush-meat crisis in West-Central Africa. *Journal of Development and Social Transformation,* 2: 99–107.

Chapter 4

Community Forestry Management and its Role in Biodiversity Conservation in Nepal

Anup K.C.

Additional information is available at the end of the chapter

http://dx.doi.org/10.5772/65926

Abstract

Community forest management is one of the successful stories of green economy sectors in Nepal recognized by the United Nation Environment Programme. It was initiated in Nepal to mitigate increasing deforestation and forest degradation and address the negative impacts on rural livelihoods. Different studies are conducted by researchers to assess the role of community forest in biodiversity conservation in Nepal. Researchers focused on analysis of biomass, carbon stock analysis, calculation of biodiversity index, change in land use and land cover, spatial analysis of forest resources, camera trapping of wild fauna and socioeconomic analysis by using different primary and secondary data collection techniques. It can be concluded that community forestry management had a great role in biodiversity conservation in Nepal. Biomass, carbon stock, growing stock, soil organic carbon, forest cover, forest products and benefit from forest resource had increased due to community forestry management. Wild animals such as leopard, porcupine, monkey and other birds were increased in the forest. It is recommended to provide skill development trainings and financial support for the installation of renewable and alternative energy technologies to minimize the use of forest resources. More researches on assessing role of community forestry management in biodiversity conservation should be conducted.

Keywords: community forestry management, conservation of flora, conservation of fauna, Nepal

1. Introduction

Forestry means use of forests for achieving specific objective that introduces it into different types [1]. The aim of industrial forestry is to produce wood-based products for national and international markets. But the objective of other approaches of forestry is to create and enhance wildlife habitat and water quality [2]. Managing forests with the express intent of benefiting neighbouring communities is community forestry (CF) [3]. Beneficial functions of the forest had attracted various actors and stakeholders, including the state, private enterprises and local forest users. Also, they have built up distinct interactions with the forest to satisfy their economic, political and social needs [4].

In community forestry, forest user group (FUG) controls and manages the local forests [1]. Harvesting and pricing of all forest products and forest management are governed by an executive committee elected in the FUG assembly [5–7]. Local people gain membership and receive cash subsidy as an incentive for forest management after registration of FUG in District Forest Office (DFO). Surplus income of community forestry forest user group (CFUG) has been used for the purpose of infrastructure development [1, 5]. Therefore, co-operation and collective actions will be obtained by transferring authority and responsibility for forest management to local users [5]. Incentives are made to control the forest through the practice of sustainable activities for income generation.

Figure 1. CFUG members carrying firewood from CF. (Source: Author).

Many rural communities that depend on nearby forests take community forestry (CF) as a tool for the globalization of the economy. It provides benefit from timber and non-timber forest resources, as shown in **Figure 1** and **Figure 2**, and also creates job opportunities. Community forestry provides benefits to local users from nearby forests. As neighbouring communities suffer most from resource degradation, community forestry provides overall role to local people in forest decision making [2]. It provides great opportunity for resource managers to utilize the indigenous knowledge of local people for sustainable management. Rural development practitioners utilize potential of forestry for forest protection, community strengthening and economic development [2].

Figure 2. Local fruits extracted from CF of Syangja. (Source: Author).

In community forestry, forest can be sustainably managed to protect natural resource and forest ecosystem functions and also provide income opportunities to community residents from traditional and non-traditional products and services. The benefits and services of forest ecosystems include non-timber forest products, watershed protection, recreational use, tourism, carbon storage, spiritual and cultural significance, genetic resources, medicinal plants and wildlife habitat [2]. In spite of the market values from timber and wood products, non-market values include environmental stability, quality of life and the economic strength of a region. Community forestry is practiced on public forest lands with the partnerships and coordination between communities and forest landowners to foster forest stewardship and economic development. It should be under the management of the local community to emphasize collaborative and participatory management in local needs and local knowledge.

1.1. Community forestry in Nepal

Historical experience shows that unless people are given user rights and ownership to control and make decisions, people lose interest in active practices of forest management [5]. Community forestry management (CFM) originates in Nepal due to the progressive degradation of hill forests caused by institutional failure [3]. Before the 1950s, forests in the Middle hills of Nepal were managed by local landlords and there was free access to non-commercial forest products. During 1951–1961, forests were nationalized and controlled by the Department of Forest (DoF). But, they were unable to manage it which creates an open access situation and local users lacked incentives to regulate forest use. It leads to unregulated extraction by creating conflicts between villagers and DoF staff. After this, land registration processes started in Nepal, which lead to encroachment and forest degradation by threatening the sustainability of livelihoods in the Middle hills [8]. There was increasing loss of forest areas due to the increase in the values of timber and other natural resources [4]. Involving local people in forest management was necessary so community forestry management was introduced to establish community-based organizations for collective management of forest resources [8].

Forest policies have been changed as the state and local communities experience forest losses and degradation after the 1950s. The process of nationalization formulates a centrally designed and scientifically informed forest policy in the context of Nepal. The new policy could not work longer as it had limited the use of forest resources and incentives for sustainable use and co-operative management. After the failure of the governmental forest management system and the revival of common-based management systems, local communities have taken back these rights to use and manage their forests and formed institutions called forest user groups (FUGs). Therefore, in the 1970s, local users' participation in forest management was reconsidered by the government after recognizing the effectiveness and benefits of common property management [9].

Conference organized by the Department of Forest in 1975 focused on the role of community in forest management. It helps in the emergence of community forest management till 1978 and further development occurs till 1993 [8]. Community forestry management was initiated on an experimental basis in the 1980s which decides to provide power and authority over resource use to the community level and return property rights to communities. Projects were initiated by the governmental institution with the support of policymakers, field staff of the forest department and project staff of the national community forestry workshop. Eventually, community forestry was legally implemented with the 1993 Forest Act and the 1995 Forest Rules with the support of local users and forestry staffs [10]. The responsibility of protection of CF is taken by local forest user groups while forestry staff plays the role of supervision [10].

The need of community involvement was identified in the National Forestry Plan after the deterioration of hill forests. After the National Forestry Plan, two amendments were made in Forest Act in 1977 and 1978, and the handover of forests has started gradually. World Bank, Australia and Britain also need changes in the forestry sector in the 1980s, and community forestry management was introduced in various policies. Decentralization Act in 1982 focused on forming the community forestry management committees for forest management, and the concept of forest user groups was introduced. To provide subsistence needs of people, the

Seventh Periodic Plan (1985–1990) gave priority to people's participation in forest management. As a second major milestone, the master plan for the forestry sector declared to handover forests in the Middle hills to FUGs by following the outcomes of first the National Community Forestry Workshop in 1987. For the reorientation of DoF staff towards this new priority, 47% of investment within the forest sector was invested in support of community forestry programmes for the new role of facilitation. Democracy in 1990 helps to form FUGs the important unit of community forestry management and a strong independent legal institution [8].

Analysts have delineated three phases of forestry development in Nepal–privatization (before 1957), nationalization (1957 to the late 1970s) and decentralization (the late 1970s onwards) [11]. Before the state took control of forests in late 1950s, most forests in rural Nepal were controlled and managed by local communities. When the government identified the need of active co-operation of local forest-dependent citizens, participation of local people in the forest management began in the late 1970s. In the history of Nepal, state was controlled by the Shah or Rana families before the democracy in the 1950s, the 1990s and after 2006. The control of forest resource and economic surplus flowed from general people to the ruling elites [11].

Before the establishment of community forestry, government of Nepal assumed that transferring forests from private groups to the state would enhance people's access to forest resources but the state imposed regulations to exclude people from controlling forest resources [11]. To mitigate deforestation and forest degradation and to address the negative impacts on rural livelihoods, community forestry management (CFM) was established as an important forestry policy in the late 1970s. It plays an important role in forest management by linking agriculture, livestock rearing and the forest [12]. It focuses on avoiding deforestation and forest degradation by implementing protective measures [13]. Involvement of local people in forest protection and management became an important policy in the forestry sector in the hilly region due to the failure of states to mitigate deforestation and forest degradation [13].

CFM has been promoted as an important step in common property resource management in Nepal [14]. To mitigate the growing deforestation and deterioration of the forest, government of Nepal made a policy based on the 1976 National Forestry Plan to involve local communities in forest management [8]. Many communities in developing countries are successful in transforming natural forests from the deteriorating state to the sustainable state. Community-based forest management is an approach to mitigate increasing deforestation and forest degradation to address the negative impacts on rural livelihoods. In Asia, this management approach quickly became widespread in different forms of community involvement in forest management and protection [8]. China Collective Forest, India Joint Forest, Philippines Community-based Forest and Nepal Community Forest are some of the examples of community managed forest. In a time duration of more than 10 years, CFM had a great role in forest conservation. Management of forest by local users and supervision by local forestry staff make CFM more successful, which is a successful example of decentralization and empowerment of local people [8].

A group of households wishing to form a CFUG should prepare an operational plan under the provisions of the Forest Act of 1993 and submit it for registration at the local District Forest Office (DFO). To prepare an operational plan for forest management, CFUG should take technical assistance from forest officials and non-governmental organizations (NGOs). There is no any legal limit for the area and size of the CFUGs as it depends on the willingness and ability of the community to manage a forest. CFUG may include all members of a village, a selected group of households, people from different village and district without any administrative boundaries. It is inclusive of households in the village and all households of one or more villages become members of a CFUG, representing diverse interest in forests. CFUG is a perpetually self-governed institution with rights to manage and fix the prices of forest products. Forest use is not restricted in legal framework and practice, but CFUGs have to pay taxes to the government for selling any forest products outside CFUG. The state retains ownership of forests but communities hold the rights to use the forests and make management decisions. With the help of operational plan, CFUGs set the price of various products, collect revenue and mobilize income for community development activities [11].

Figure 3. General meeting of CFUG members. (Source: Author).

CFUG members can participate in decision-making process through role-based meetings, executive committee involvement, annual assemblies and forest management plans formulation as shown in **Figure 3** and **Figure 4**. Each CFUG prepares its own constitution and operational plan, registers and approves from DFO, defines the social arrangements, responsibilities and rights of the group and makes arrangement for forest management. The strategy, constitution and operational plan are prepared by following standard guidelines and norms but varies from group to group to adapt local traditions and practices. Each CFUG elects a specified number of members to an executive committee for a period of 1–3 years to carry out day-to-day decisions about forest management [11].

Figure 4. Checklist and feedback collection from CFUG. (Source: Author).

Community forestry management in Nepal is successful in providing important rights, including decision making, empowerment over forest management and use, and access to forest resources at the community level. Community forestry policy and institutional innovations contribute to improved welfare and livelihood security in Nepal by increasing access of CFUG to forest products and by providing positive impacts on income, employment and entrepreneurial opportunities, livelihood diversification, and broader community development activities. The role of CFM to overall livelihood security is critically important in Nepal because more than 70% of Nepal's population depends on agricultural livelihoods that encompass complex interactions between agriculture, forestry and livestock systems [11].

Up to date, a total of 1,798,733 ha of community forest is handed over to 18,960 community forest user group throughout the country [15]. The trend of conversion of public forest into community forest is increasing rapidly with the need and interest of local community in conserving forest.

1.2. Role of CFM in biodiversity conservation

Different life forms or varieties of life are called biodiversity, and care and management of biological materials are called biodiversity conservation [16]. It is categorized as species diversity, ecological diversity and genetic diversity [17]. Due to the unique geographical location with diverse climate and altitude, Nepal has great diversity of flora and fauna [16, 17]. Nepal consists of 0.1% of the terrestrial area of the earth with 118 ecosystems, 75 vegetation types and 35 forest types [17]. It consists of 5000 species of flowering plants, 2252 species of moths, 635 species of butterflies, 185 species of fishes, 844 species of birds and 181 species of mammals. For the conservation of biodiversity, there is provision of protected areas, zoo, different types of law, conventions, non-governmental organizations (NGOs), local and national authorities and national and international organizations [16].

Figure 5. Scenic beauty of forestry combined with water. (Source: Author).

Community forestry is successful in decreasing resource degradation and helpful in the conservation of biodiversity [18]. Implementation of community forest management has improved the forest condition and biodiversity in the hills of Nepal as compared to degraded forest in the past. It could be a suitable option to conserve biodiversity, but it focuses on sustainable forest product and keeping biodiversity conservation in less priority. Its aim is to supply forest products to local users rather than to conserve biodiversity [19]. There is a considerable role of community forestry in biodiversity conservation of Nepal. The impacts on biodiversity of plant species are clear but it is less obvious in the case of faunal biodiversity. Community forestry had protected or re-established habitat and helpful in the survival of birds and animals. Operational plans also include prohibitions against hunting at the request of local people [20]. For the conservation of forest and its biodiversity, CFUGs are voluntarily involved in fencing, planting and meetings. It is helping in carbon sequestration and increasing the forest cover by controlling deforestation and forest degradation [1].

Various studies have demonstrated a significant increase in forest condition under community forestry showing that it is a proven model for controlling deforestation and forest degradation.

CF helps in supporting livelihood in hilly area by providing necessary forest products, such as fodder, firewood, timber, leaf litter and agricultural tools. CFM also helps in conservation of flora and fauna. There is a growing concern that CF is prioritizing only towards sustainable management of forest resources and less towards biodiversity conservation. The aim of community forestry is to supply forest products to local users rather than to conserve biodiversity. Currently, there is evidence that CFUGs are slowly moving towards active forest management. Effective management of CF leads to sustainable production and sustainable harvest of forest resource. Sustainable harvest of forest resource helps to fulfil forest product needs and also helps in livelihood enhancement of local people [18].

CF has been successful to provide forest resource need of people by enhancing the forest cover. Forest enhancement is increasing ecological services of forest, water resource management, biodiversity conservation, carbon stock, greenery enhancement and air quality management, as shown in **Figure 5** [12]. In actual fact, CFM provides win-win situation in atmospheric carbon dioxide mitigation and biodiversity conservation in global scale; and livelihood enhancement and greenery enhancement in local scale are shown in **Figure 6**. Community forestry also had co-benefits of reducing poverty, addressing social exclusion and creating rural employment [21].

Figure 6. Agroforestry system practiced in CF area. (Source: Author).

In Nepal, local communities have come a long way in conserving forest ecosystems and nurturing local institutions for democracy and social justice. The historical context for the emergence of community forestry in Nepal dates back to the 1950s, when the Government of Nepal nationalized all the forests hoping to optimize the use of natural resources and conserve it sustainably. Communities were totally excluded from the forest management process threatening the livelihood of the rural people. This exclusion led to massive deforestation and degradation of natural resources solely because the community viewed the state as an enemy causing destruction of forest. The situation of environmental crisis had emerged due to lack of participation of community in management of forests. By the late 1970s, Nepal had lost almost 2.2 million hectares of forest cover resulting in serious downstream flooding. In the Kavre and Sindhupalchok districts of central Nepal, a study found that shrub land and grass land have been converted into productive forests increasing the forest area from 7677 to 9678 ha [11, 22]. Three different studies conducted in mountain ecosystem for a time period of 25 years (1976–1989–2000) showed that forest cover had increased as compared to the past. Due to the increase in forest cover, small patches were merged into larger ones decreasing their number (from 395 to 175) and increasing forest area (794 ha). Thus, there is an overall improvement in forest protection contributing to local environmental conservation and increased greenery [11, 23].

2. Methodology

Different research studies are conducted by different researchers to assess the role of community forest in biodiversity conservation in Nepal by applying different methodologies. Issues raised by researchers in their research studies were reviewed to find out tools and techniques applied by them. It would support the researchers to identify the subject of research with appropriate tools and techniques. It would make the new researchers easy and simple to select appropriate literature necessary for them. This chapter provides the location, aim and methodology of different research studies throughout the world.

With an objective to examine the impacts of forest management on biodiversity in Nepal, Acharya [19] conducted a study on two CFUGs in the Mid-hill region of Parbat district in Nepal. The study area was selected on the basis of similar socio-economic, ecological conditions, area, forest types and biophysical factors. Forest biodiversity information was collected using six transects walk at three different altitudes in the east-west and north-south directions in each of the CFUGs with the help of informal interviews with CFUG members by applying tools and techniques of participatory rural appraisal (PRA).

Thoms [24] conducted a study to examine whether community forestry is elite dominated and not successful in livelihoods improvement of CFUGs. For this purpose, primary data were collected from 6 months of field research between October 2002 and April 2003 in four hill districts and two Terai districts. Data were collected from 2871 household surveys selected through multi-stage area probability sampling.

With an aim to compare land use changes between village development committees (VDCs) with and without community forests, Gautam et al. [25] conducted a study in the Roshi watershed of Kabhrepalanchok district in the Middle Hills of Nepal. Spatial analysis was based on two land use data sets, 1978 data compiled by the land resource mapping project (LRMP) and 1992 data compiled by Survey Department of His Majesty's Government of Nepal (HMGN).

With an aim to evaluate forest condition in community forests, national forests and protected areas in the Nepal Terai, Nagendra [26] conducted a study in three International Forestry Resources and Institutions (IFRI) research sites in the Chitwan district of Nepal. The site was selected to cover the east-west range at an altitude of 195–425 m above the sea level in *Shorea robusta* dominated tropical moist deciduous hardwood forests. Assessment of forest condition was carried out through the use of forest plots evaluation by a professional forester and interviews with the local communities depending on forest.

Figure 7. Measurement of diameter of tree inside CF. (Source: Author).

Adhikari et al. [18] studied the relationship between key household characteristics and common property resources in eight community forest of two districts, Kabhre Palanchok and Sindhu Palchowk, in the mid-hills of Nepal. The study was based on information collected through a household survey. A total of 20% stratified sample of households from each income group was chosen by compiling a census of village households with participatory rural appraisal (PRA) techniques.

With an aim to assess success of restoration in community forest using a reference of semi-protected natural forest, Baral and Katzensteiner [27] conducted a study in CF and better protected municipality owned forest (MF) in similar topographic positions in Dhulikhel of Kavrepalanchowk district. The diversity of vascular plants and forest structure was compared with the help of primary data of tree height, diameter at breast height (DBH) and crown width with the help of transect survey, clinometers and diameter tape.

To assess improved condition of forests by collective action of local communities, Shrestha and McManus [28] conducted a study in three CFUGs of Nepal. Data collection was directly carried out by rapid forest assessment (RFA), household questionnaire interview (HQI), group discussion, participant observation and informal talks. Rectangular plots were established in each community forest of size 100 m^2 (10 m 10 m) to capture plantation in a recently harvested site and dense forest with mature trees and to represent the diversity of forests within the sample plots.

To identify the role of community forests in the conservation of faunal diversity of Satbariya Range Post of Dang district, Pokhrel and Shah [29] conducted a study with the help of questionnaire survey, group discussion and line transect methods. They collected data of faunal diversity, abundance and distribution pattern of the wild animals and wildlife-people conflict.

To estimate the climate change mitigation potential from carbon stock of the forest, K. C. et al. [14] conducted a study in Ghwangkhola Sapaude Babiyabhir Community Forest (GSBCF) in Syangja district of Nepal. Their study was based on carbon stock measurement and review of past studies.

Figure 8. Measurement of height of tree inside CF. (Source: Author).

With a special focus to study the impact of forest resource use on carbon stock of forest, Paudel and K. C. [30] conducted a study in Kafle Community Forest of Lalitpur district in Nepal. To conduct carbon stock measurement, focus group discussion and key informant interview, field visit was conducted in different time of year in 2012 and 2014. Biomass measurement was conducted directly in the field for trees and sapling by following national guideline as shown in **Figure 7** and **Figure 8**. Twenty composite samples of leaf litter, herbs, grasses and soil collected in the field were brought to the laboratory for detailed analysis of biomass and carbon stock.

With a special focus to study the feasibility of community forest management, K. C. and Manandhar [31] carried out research on GSBCF of Syangja district in Nepal. Their study was based on carbon stock measurement, household survey, focus group discussion, key informant interview and review of past studies.

3. Role of CFM in conservation of flora

As stated in the introductory section, CFM had a great role in the conservation of flora and fauna. To identify the status of community forest, findings of different research studies are documented below.

Gautam [32] conducted a study to assess the carbon sequestration rate of the agroforestry system, natural forest and annual cropping system in the Terai region of Nepal. Natural forest had the highest carbon stock of 98 ton/ha. Carbon stock in the annual cropping system ranges from 33.2 to 55.5 ton/ha while that of orchard plantation ranges from 35 to 74.6 ton/ha. Similarly, soil organic carbon (SOC) in natural forest, vegetable field and streamside were 53.2, 52.6 and 3.6 ton/ha, respectively.

Acharya [19] observed that active management by CFUGs contradicts with biodiversity conservation. Forest types are slowly converting to monoculture from mixed, shrub and tree diversity is decreasing gradually, and shrub land areas are gradually converting to high forest land. Active forest management favouring specific useful plant may introduce more homogeneity into the forest structure with consequent loss of biodiversity. It will lead to the modification of forest types and ecosystem in the mid-hills of Nepal affecting ecological functions and services of forests.

Thoms [24] concluded that community forestry is quite successful in terms of forest protection and management but at the cost of the poorest households. Community forestry is fairly successful in conservation but not in improving rural livelihoods.

Gautam et al. [25] observed that VDCs having community forests before 1992 sustained less total loss of forested area (1.9%) than VDCs without community forest (9.9%). High forest area was six times higher in VDCs with community forests (77%) than the VDCs without community forests (13%). Loss of shrub land in VDC with community forest was 50% greater than that of VDCs without community forest.

Nagendra [26] observed that vegetation density and species diversity were highest in national park forest, followed by national forest and community forests. Community forests were Significantly poorer as compared to national forests in species richness and Shannon species diversity of tree; sapling density, sapling diameter, sapling richness, sapling Shannon species diversity, sapling girth and sapling height. Trees located in community forests were Significantly taller with high density than those within national forests, but there was no difference in tree size (diameter). Community forests have Significantly low species richness, low Shannon species diversity and smaller diameter saplings as compared to national park forest. There was low level of grazing, low tree lopping but proper fencing in community forest as compared to the national forest.

Adhikari [33] tried to examine the contribution of community forestry to household-level income with particular emphasis on group heterogeneity and equity in benefit distribution. The household level benefits suggest that poorer households are currently benefiting less from community forestry. But, poor are not more dependent than the rich in community forest. Regression analysis shows that socio-economic conditions and ownership of private property are directly related to revenue generated from community forest. Households having more land and livestock get more benefits from community forest. Educated people and female-dominated household get less benefit from forest resources.

Bhatta [34] conducted a study in mixed broad leaved forests of Phulchowki watershed, Lalitpur. The carbon stock in above ground in natural forest and community forest ranges from 91.89 to 112.79 and 55.30 to 67.04 ton/ha, respectively. Similarly, the carbon stock in soil in natural forest and community forest ranges from 195 to 223 and 150 to 160 ton/ha, respectively.

Shrestha and McManus [28] observed that local communities are effectively protecting the forest through direct efforts of users or through forest watchers or sometimes both. Improvement in forest condition was seen by reversing degradation and regenerating degraded areas but not ideally for biodiversity conservation. The forests have low species diversity as the trees and poles are dominated by few species promoted by the FUG for their social, economic and political values.

Dahal [35] conducted a study in Sunaulo GhampaDanda CF in Kathmandu. The biomass organic carbon in pine forest and mixed broad leaf forest was 116 ± 16.39 and 25.95 ± 8.09 ton/ha, respectively. The soil organic carbon in pine forest and mixed broad leaf forest was 10.12 ± 1.03 and 24.62 ± 1.18 ton/ha, respectively. The carbon sequestration status in pine forest and mixed broad leaf forest was 1 and 2.95 ton/ha per year, respectively. The additional benefit to CFUG by carbon trading was $ 563.15 per annum.

Karky [36] conducted a study in three community forest of Manang, Lalitpur and Ilam district in Nepal. From measurement of carbon stock, it was observed that the carbon stock of community forest with SOC up to 1 m depth (without leaf litter, herbs and shrubs) was 138 ton/ha or 504 ton CO_2/ha in three districts of Nepal. He found that the annual incremental rate for carbon sequestration in forest under CFM was 1.92 and 7.04 ton/ha per year excluding soil organic carbon. He also found that when CFUGs are permitted to use forest resource, the breakeven price for per ton CO_2 is $0.55 for Illam, $3.70 for Lamatar and $2.30 for Manang.

Baral and Katzensteiner [27] observed that maximum tree height (13.5 m) and the maximum DBH (29.5 cm) were observed in managed forest (MF). Trees with higher diameters have a higher basal area in MF, but 5–15 cm DBH trees have higher a basal area in CF. CF management activities have affected plant community composition, species richness and distribution, and age class distribution of the trees. CF was less diverse with uniform stands of tree species compared to MF. Overall diversity of vascular plants was maintained by providing proper niches for rich under storey vegetation.

Gurung [37] conducted a pilot study in western Terai and had estimated the average forest carbon stock to be around 231 ton/ha. The carbon stock in trees above ground, below ground and in soil organic carbon (SOC) had been estimated to be about 68, 18, and 143 ton/ha, respectively. This clearly indicates that the share of SOC was almost 60% of the total forest carbon stock.

Thagunna [38] conducted a study in Bailbanda Buffer zone CF, Kanchanpur. The total carbon stock of CF was 78.46 ton/ha. The benefit from carbon trade was $ 57,640 at the rate of $ 12.5/ton C.

Aryal [39] conducted a study in Toudol Chhap CF, Sipadol, Bhaktapur. The total carbon content of pine forest and mixed broad leaf forest were 167.04 and 101.91 ton/ha, respectively.

Bhusal [40] conducted a study in Nagmati watershed in Shivapuri National Park. The SOC and total carbon content in the sampled area (14 ha) were found to be 9782.11 ± 25.18 ton/ha corresponding to a total of 167442.26 ± 42076.82 ton carbon content in the Nagmati watershed (1406 ha). The total carbon content of Shivapuri National Park (5860.8 ha, i.e. 40% of the total area of park which is forest) excluding soil was 699961.20 ± 175894.32 ton.

Dhakal [41] measured the total carbon stock in Pashupati Community Forest, Sarlahi district of Janakpur zone. The total carbon stock was found to be higher in naturally regenerated forest i.e. 181.83 ± 26.34 ton/ha followed by planted forest with 159.49 ± 31.96 ton/ha. The recent amount of total carbon stock of 133.65 ± 37.05 ton/ha was found in enriched forest.

ICIMOD, ANSAB and FECOFUN (2010) had performed baseline study in 104 community forests (CF) of three watershed areas of Nepal; Kayarkhola of Chitwan district, Charnawati of Dolakha district and Ludhikhola of Gorkha district. Analysis of the DBH distributions of all strata follows a left-skewed trend, indicating most of the trees in all the strata were younger, and there was potential to enhance forest carbon stock by encouraging tree growth. Forest carbon stock in dense and sparse strata of Kayarkhola, Charnawati and Ludikhola watershed were 296.44 and 256.70, 228.56 and 166.75, 216.26 and 162.98 ton/ha, respectively.

Mishra [42] conducted a study in Chapako CF, Kathmandu. The biomass carbon and soil organic carbon (SOC) of CF were 119.742 and 32.29 ton/ha, respectively. There was potential of storing and sequestering carbon in the CF.

Community forest user groups are giving less attention to biodiversity, ecosystem functions and services due to short-term economic motive, elite sanction and knowledge gap. They are unaware about maintaining biodiversity, ecosystem services and sustainable forest management. Monoculture of high economic valuable species and greenery of the forest are prioritized rather than the natural forest. Seedling plantation, wildlife hunting control and regulating forest encroachment assist biodiversity conservation but species selection, removal of unwanted species and traditional knowledge depletion have negative impact on biological diversity [43].

The study reveals that the carbon stored in the forest soil is almost double than the biomass carbon [44]. The biomass in the Gwangkhola Sapaude Babiyabhir community forest, Syangja, Nepal, was found to be 164 ton/ha, with yearly increment of 0.95 ton/ha. The total carbon stock of the forest was 122.29 ton/ha, including soil organic carbon and below ground carbon of 45.18 and 12.85 ton/ha, respectively. The forest was dominated by *Schima wallichi, Castanopsis indica* and *Pinus roxburghii* [45].

Figure 9. Increase in forest cover and leaf litter in CF. (Source: Author).

K. C. et al. [14] measured the biomass in above ground shoot and below ground root of trees, shrubs, leaf litter, herbs and grass (LHG) in community forest of Syangja district in Nepal. It was observed that above ground biomass of trees was highest (126.3 ton/ha) followed by below ground biomass (27.34 ton/ha), sapling biomass (2.88 ton/ha) and leaf litter, herbs and grass biomass (7.54 ton/ha). Carbon stock in forest (122.29 ton/ha) was increasing at the rate of 0.45 ton/ha per year. Atmospheric carbon dioxide (CO_2) was mitigated by the forest at the rate of 1.64 ton/ha per year. The species diversity of *Schima wallicchi* was highest followed by *Castanopsis indica*.

K. C. and Manandhar [31] observed the total carbon stock of 155.04 ton/ha with soil organic carbon of 50.15 ton/ha in the forest. The more number of trees below 20 cm DBH shows that the forest is conserved after handing it to CFUG and newly grown plants are increasing thereafter. Above ground tree carbon had increased from 59.36 to 80.09 ton/ha while soil organic carbon had increased from 45.18 to 50.15 ton/ha from 2011 to 2014. The carbon stock of *Schima wallichi, Castanopsis indica* and *Pinus roxburgi* was gradually increasing from 2011 to 2014.

Paudel and K. C. [30] observed that community forest management had helped in conservation of plants and animals as forest is getting denser than past, as shown in **Figure 9** and **Figure 10**. Carbon stock in all forms of plants as measured in 2014 was higher than that of 2012 with an annual carbon sequestration rate of 1.52 ton/ha. The forest was dominated by *Schima wallichi, Castanopsis indica, Alnus nepalensis* and *Pinus roxburgi*.

Figure 10. Researcher conducting study inside dense CF. (Source: Author).

4. Role of CFM in conservation of fauna

There are very few research studies conducted to assess the role of CFM in fauna conservation of Nepal. Some of the research studies conducted on the concerned topics were reviewed and documented below.

Paudel and K. C. [30] observed that carbon stock in all forms of plants as measured in 2014 was higher than that of 2012 with an annual carbon sequestration rate of 1.52 ton/ha. Community forest management had helped in conservation of plants and animals. Wild animals such as leopard, porcupine, monkey and other birds were increased in the forest and were frequently seen nearby the forest destroying the crops of people. The forest is becoming denser than past according to the view of local people.

To find the condition of animals in the Setidevi community forest and Gyaneshwar community forest, camera trapping technology was used. In the forest area of 500 ha, 181 animal species including one-horned rhino, Royal Bengal tiger and python have been observed. Among these 125 bird species and 19 mammals have been spotted [46].

Pokhrel and Shah [29] observed the increased frequency and movement of wild elephant and blue bull due to the establishment of community forests. Twenty-five mammals, 16 herpetofauna and 163 bird species were recorded in their study area indicating availability of suitable habitat for the species. In the study of 10 transects, they encountered 251 different signs of the wild fauna. Local people had suffered from economic loss of crop damage and livestock due to the increasing number of wildlife in the community forest.

5. Conclusions

Community forest management is an approach to mitigate increasing deforestation and forest degradation to address the negative impacts on rural livelihoods. Studies have demonstrated a significant increase in forest condition under community forestry showing that it is a proven model for controlling deforestation and forest degradation. It has co-benefits of reducing poverty and addressing social exclusion by creating rural employment. It is contributing to livelihood promotion such as fulfilling the basic needs of local communities investing money in supporting income generation activities of the poor people and providing access to the forestland for additional income or employment.

Different research studies are conducted by different researchers to assess the role of community forest in biodiversity conservation in different study areas of Nepal by applying different methodologies. Researchers had focused on analysis of biomass, carbon stock analysis, calculation of biodiversity index, change in land use and land cover, spatial analysis of forest resources, camera trapping of wild fauna and socioeconomic analysis by using different primary and secondary data collection techniques. They are using national guideline and their own derived methodologies for assessing biomass, carbon stock, measurement of biodiversity index and analysis of flora and fauna.

It was concluded that community forestry management had a great role in biodiversity conservation in Nepal. Biomass, carbon stock, growing stock, soil organic carbon, forest cover, forest products and benefit from forest resource had increased due to CFM as compared to past. The number and density of trees of highly productive plant had increased while the number and density of less productive shrubs and bushes had decreased. Forest biomass and carbon in different form of plants, above ground tree biomass, above ground sapling biomass, leaf litter herbs and trees and underground biomass had increased gradually after CFM implementation. Wild animals such as leopard, porcupine, monkey and other birds were increased in the forest and were frequently seen nearby the forest destroying the crops of people. The forest is getting denser and providing habitat to the wild animals as compared to past according to the view of local people.

As community forestry management had great role in biodiversity conservation of Nepal, there is a need of more funding for its sustainable management. Local people are working hard and devoting their time voluntarily for sustainable harvest of forest resource and conservation of flora and fauna. If they do not get adequate benefit of forest resource and monetary benefit from job employment and other income-generating activities, they will start using forest products for sustaining their livelihood and fulfilling their day-to-day need. It would cause utilization of more forest resource and decrease in biodiversity of plants and animals. It is recommended to provide skill development trainings, income-generating activities, high yield forest resource and non-timber forest products and also provide financial support for the installation of renewable and alternative energy technologies to minimize the use of forest resources. In addition, more research studies on assessing the role of CFM in biodiversity conservation should be carried out to find out the feasibility of CFM in the Nepalese context for biodiversity conservation.

Author details

Anup K.C.

Address all correspondence to: kcanup04@gmail.com

Nepal Electricity Authority, Environment and Social Studies Department, Kharipati, Bhaktapur, Kathmandu, Nepal

References

[1] Anup, K.C., I. Koirala, and N. Adhikari, *Cost Benefit Analysis of a Community Forest in Nepal*. Journal of Sustainable Forestry, 2015. 34(3): 199–213.

[2] Brendler, T. and H. Carey, *Community Forestry, Defined*. Journal of Forestry, 1998. 96(3): 21–23.

[3] Anup, K.C., *Community Forest Management: A Success Story of Green Economy in Nepal*. Journal of Environmental Science, 2016. 2: 148–154.

[4] Hobley, M. and Y. Malla, *From Forests to Forestry. The Three Ages of Forestry in Nepal: Privatisation, Nationalisation and Populism*, in *Participatory Forestry: The Process of Change in India and Nepal, Rural Development Forestry Study Guide 3*, M. Hobley, Editor. 1996, Rural Development Forestry Network Overseas Development Institute: London.

[5] Gilmour, D.A. and R.J. Fisher, *Evolution in Community Forestry: Contesting Forest Resources*. 1998, RECOFTC: Bangkok, Thailand.

[6] Khanal Chhetri, B.B., J.F. Lund, and O.J. Nielsen, *The Public Finance Potential of Community Forestry in Nepal*. Ecological Economics, 2012. 73(2012): 113–121.

[7] Anup, K.C., et al., *Increase of Forest Carbon Biomass due to Community Forestry Management in Nepal*. Journal of Forestry Research, 2016.

[8] Springate-Baginski, O., et al., *Community Forest Management in the Middle Hills of Nepal: The Changing Context*. Journal of Forest and Livelihood, 2003. 3(1): 5–20.

[9] Brown, D., et al., *From Supervising Subjects to Supporting Citizens: Recent Developments in Community Forestry in Asia and Africa*, in *Natural Resource Perspective*. 2002, Overseas Development Institute: London.

[10] Pokharal, B.K., *Livelihoods, Economic Opportunities and Equity: Community Forestry and People's Livelihoods*. Journal of Forestry and Livelihood, 2001. 1(1): 16–18.

[11] Ojha, H., L. Persha, and A. Chhatre, *Community Forestry in Nepal, A Policy Innovation for Local Livelihoods*, IFPRI Discussion Paper. 2009, International Food Policy Research Institute: Washington.

[12] Karky, B.S. and K. Banskota, *The Kyoto Protocol and Community-Managed Forests*, in *Reducing Carbon Emissions through Community-Managed Forests in the Himalayas*, K. Banskota, B.S. Karky, and M. Skutch, Editors. 2007, International Centre for Integrated Mountain Development (ICIMOD): Kathmandu, Nepal. pp. 23–37.

[13] Banskota, K., B.S. Karky, and M. Skutch, *Reducing Carbon Emissions through Community-Managed Forests in the Himalayas*. 2007, ICIMOD: Kathmandu, Nepal.

[14] Anup, K.C., et al., *Climate Change Mitigation Potential from Carbon Sequestration of Community Forest in Mid Hill Region of Nepal*. International Journal of Environmental Protection, 2013. 3(7): 33–40.

[15] DoF, *CFUG Database Record Available in MIS*. 2015, Community Forestry Division, Department of Forest, Kathmandu, Nepal: Kathmandu.

[16] Thapa, D., *Biodiversity Conservation in Nepal*. Bibechana, 2010. 6: 31–36.

[17] Shakya, R., et al., *Biodiversity Conservation*. Banko Janakari, 2009. 19(1): 1–2.

[18] Adhikari, B., S.D. Falco, and J.C. Lovett, *Household Characteristics and Forest Dependency: Evidence from Common Property Forest Management in Nepal*. Ecological Economics, 2004. 48: 245–257.

[19] Acharya, K.P., *Does Community Forests Management Supports Biodiversity Conservation? Evidences from Two Community Forests from the Mid Hills of Nepal*. Journal of Forest and Livelihood, 2004. 4(1): 44–54.

[20] Fisher, R.J., *Collaborative Management of Forests for Conservation and Development*, Issues in Forest Conservation. 1995, IUCN-The World Conservation Union, World Wide Fund for Nature. pp. 1–17.

[21] Anup, K.C., *Feasibility Analysis of REDD+ A Case Study in Ghwangkhola Sapaude Babiyabhir Community Forest of Syangja, Nepal*, in *Central Department of Environmental Science*. 2012, Tribhuvan University: Kathmandu.

[22] Jackson, W.J., et al., *Land-Use Changes in Two Middle Hills Districts of Nepal*. Mountain Research and Development, 1998. 18(3): 193–212.

[23] Gautam, A.P., et al., *Land Use Dynamics and Landscape Change Pattern in a Mountain Watershed in Nepal*. Agriculture, Ecosystems and Environment, 2003. 99(1–3): 83–96.

[24] Thoms, C.A., *Community Control of Resources and the Challenge of Improving Local Livelihoods: A critical Examination of Community Forestry in Nepal*. Geoforum, 2008. 39: 1452–1465.

[25] Gautam, A.P., E.L. Webb, and A. Eiumnoh, *GIS Assessment of Land Use/Land Cover Changes Associated with Community Forestry Implementation in the Middle Hills of Nepal*. Mountain Research and Development, 2002. 22(1): 63–69.

[26] Nagendra, H., *Tenure and Forest Conditions: Community Forestry in the Nepal Terai*. Environmental Conservation, 2002. 29(4): 530–539.

[27] Baral, S.K. and K. Katzensteiner, *Diversity of Vascular Plant Communities along a Disturbance Gradient in a Central Mid-Hill Community Forest of Nepal*. Banko Janakari, 2009. 19(1): 3–10.

[28] Shrestha, K.K. and P. McManus, *Collective Action of Local Communities in Forest Conservation and Utilisation: Critical Reflections from Nepalese Community Forestry*, in *Small-Scale Forestry and Rural Development – the Intersections of Ecosystems, Economics and Society*. 2006, Galway, Ireland. pp. 458–477.

[29] Pokhrel, G.K. and K.B. Shah, *Role of Community Forests in Faunal Diversity Conservation: A Case Study of Community Forests Within Satbariya Range Post of Dang District*. Nepal Journal of Science and Technology, 2008. 9: 111–117.

[30] Paudel, R. and K.C. Anup, *Impact of Resource Utilisation and Management on Carbon Stock in the Community Forest of Nepal*. 2014, Institute of Science and Technology, Tribhuvan University: Kirtipur, Kathmandu.

[31] Anup, K.C. and R. Manandhar, *Feasibility Analysis of Community Forestry Management in Syangja, Nepal*. 2014, Tribhuvan University: Kathmandu.

[32] Gautam, K.R., *Carbon Sequestratoin on Agro Forestry and Annual Cropping System in Inner Terai, Central Himalaya*. 2002, Management of Natural Resources and Sustainable Agriculture, Agricultural University of Norway: Norway.

[33] Adhikari, B., *Property Rights and Natural Resources: Socio-Economic Heterogeneity and Distributional Implications of Common Property Resource Management*, in *Sandee Working Paper*, South Asian Network for Development and Environmental Economics (SANDEE), Editor. 2003, Kathmandu, Nepal.

[34] Bhatta, P., *Carbon Stock Capacity of Mixed Broad leaved Forests of Phulchowki Watershed, Lalitpur*, in *Central Department of Environmental Science*. 2004, Tribhuvan University: Kirtipur, Kathmandu, Nepal.

[35] Dahal, P., *Carbon Sequestration Status at Sunaulo Ghampa Danda Community Forest, Kathmandu*, in *Central Department of Environmental Science*. 2007, Tribhuvan University: Kirtipur, Kathmandu, Nepal.

[36] Karky, B.S., *The Economics of Reducing Emissions from Community Managed Forests in Nepal Himalaya*. 2008, Dissertation to Obtain the Degree of Doctorate, Centre for Clean Technology and Environmental Policy, University of Twente, Enschede: The Netherlands.

[37] Gurung, M., *Assessment of Forest Carbon Potential of Riverine Forests at the Khata Corridor and Lamahi-mahadevpuri Complex*. 2009, Presented in National Sharing Workshop in Everest Hotel on 25th August 2009. Organized by WWF Nepal/WINROCK International and Terai Arc Landscape Kathmandu, Nepal.

[38] Thagunna, L.K., *Estimation of Carbon Stock of Bailbanda Buffer Zone Community Forest, Chadani VDC-7, Kanchanpur*, in *Central Department of Environmental Science*. 2009, Tribhuvan University: Kirtipur, Kathmandu, Nepal.

[39] Aryal, C., *Status of Carbon Stock at Toudol Chhap Community Forest, Sipadol, Bhaktapur*, in *Central Department of Environmental Science*. 2010, Tribhuvan University: Kirtipur, Kathmandu, Nepal.

[40] Bhusal, R.P., *Carbon Stock Estimation of Nagmati Watershed in Shivapuri National Park*, in *Central Department of Environmental Science*. 2010, Tribhuvan University: Kirtipur, Kathmandu, Nepal.

[41] Dhakal, K., *Carbon Stock Estimation of Pashupati Community Forest*. 2010, Master Thesis, College of Applied Science, Tribhuvan University: Kathmandu, Nepal.

[42] Mishra, N., *Estimation of Carbon Stock at Chapako Community Forest*, in *Central Department of Environmental Sciences*. 2010, Tribhuvan University: Kirtipur, Kathmandu, Nepal.

[43] Shrestha, U.B., B.B. Shrestha, and S. Shrestha, *Biodiversity Conservation in Community Forests of Nepal: Rhetoric and Reality*. International Journal of Biodiversity and Conservation, 2010. 2(5): 98–104.

[44] Khadka, K.J., *Carbon Stock Estimation and Analysis of Altitudinal Carbon Content Variation along Thulonagi Community Forest, Jiri, Dolakha*, in *Central Department of Environmental Science*. 2011, Tribhuvan University: Kirtipur, Kathmandu, Nepal.

[45] Anup, K.C., *Feasibility Analysis of REDD, A Case Study in a Community Managed Forest in Syangja, Nepal*, in *Central Department of Environmental Science*. 2012, Tribhuvan University: Kirtipur, Kathmandu, Nepal.

[46] WCN. *Endangered Species Found in Chitwan Community Forests*. 2013

Chapter 5

Wildlife Tuberculosis: An Emerging Threat for Conservation in South Asia

Jeewan Thapa, Chie Nakajima, Kamal P. Gairhe,
Bhagwan Maharjan, Sarad Paudel, Yogendra Shah,
Susan K. Mikota, Gretchen E. Kaufman,
Deborah McCauley, Toshio Tsubota,
Stephen V. Gordon and Yasuhiko Suzuki

Additional information is available at the end of the chapter

http://dx.doi.org/10.5772/65798

Abstract

Wildlife tuberculosis (TB) is becoming one of the emerging challenges for conservation globally. South Asian region is home to many endangered species like Asian elephants, rhinoceros, and Bengal tigers. Although it carries more than one-third of global burden of human TB, TB in livestock and wildlife has not been adequately studied. This chapter reviews the present knowledge and information about animal-adapted members of *Mycobacterium tuberculosis* complex and wildlife TB in South Asia. Recent studies of TB from different wild animals in Nepal and Bangladesh have found that *M.orygis* is an emerging threat of wildlife TB in the region. These studies have demonstrated wide diversity of *M. orygis* strains circulating in the region indicating its endemic distribution. *M. orygis*–associated TB was discovered from a free-ranging rhinoceros in Nepal and the finding could signify threat of TB in other wild animals, including a possibility of unknown maintenance host. Recent studies also revealed an emerging challenge caused by TB to elephants in different South Asian countries like Nepal, India, and Sri Lanka. Wildlife TB is becoming a conservation challenge in South Asia, but given the paucity of research in this area, it is overlooked and underexplored.

Keywords: wildlife tuberculosis, *Mycobacterium tuberculosis* complex, *Mycobacterium orygis*, Asian elephants, South Asia

1. Introduction

South Asia or the South Asian Association for Regional Cooperation (SAARC) region consists of eight countries, namely, Afghanistan, Bangladesh, Bhutan, India, Maldives, Nepal, Pakistan, and Sri Lanka. This region is one of the world's hotspots for biodiversity as it includes 17 regions in the World Wildlife Fund (WWF)'s Global-200 biodiverse ecoregions [1], and 2 of 25 priority regions for biodiversity conservation [2]. It is also considered as a high-risk region for emerging infectious diseases that could originate in wildlife [3].

The *Mycobacterium tuberculosis* complex (MTBC) is a group of genetically closely related pathogens that can cause tuberculosis (TB) in humans and animals [4]. The MTBC includes the typical human-adapted pathogens *M. tuberculosis* and *M. africanum* [5] and those members reported to cause TB in a range of animals which are *M. bovis, M. caprae, M. pinnipedii, M. microti, M. orygis, M. suricattae,* Dassie bacillus, *M. mungi,* and so-called the Chimpanzee bacillus (**Figure 1**) [6–14]. *M. bovis,* the most studied animal TB pathogen, is now considered as an emerging pathogen of free-ranging wildlife and an emerging threat to several protected wildlife species [15]. Similarly, the threat of TB in endangered species, such as chimpanzees in the Ivory Coast [14] caused by the chimpanzee bacillus, and to free-ranging rhinoceroses in Nepal by *M. orygis* [16], are examples of the challenges that TB poses for wildlife conservation. Through this review, we will provide a global overview of animal-adapted members of the MTBC; highlight their importance to defining "One Health" connections between humans, animals, and wildlife; and draw attention to their emerging threat to wildlife conservation.

TB caused by *M. tuberculosis* is a high-priority disease in South Asia as it carries 34% of the global TB burden, with 3 million people infected and a mortality of 0.4 million people in 2013 [17]. However, TB caused by *M. bovis* is a neglected disease both in the livestock sector and in human health in the region. In human health, MTBC species differentiation is not a priority because it requires mycobacterial culture and subsequent use of specialized molecular tests for diagnosis [18]. There are only a few confirmatory reports of zoonotic TB caused by *M. bovis* in the region [19] and we are not aware of any livestock TB control programs. In addition, despite the limited number of publications that describe TB in different wildlife species, to the best of our knowledge, there are no reports of confirmed diagnosis of *M. bovis* infection from wildlife in the region. Thus, zoonotic TB caused by *M. bovis* and other members of the MTBC that may be present in livestock, wildlife, as well as humans is largely overlooked. In recent studies we have demonstrated the conservation challenge posed by *M. orygis* in Nepal [16, 20]. Complementing this finding, from our studies in Bangladesh [21], we have demonstrated a wide distribution of *M. orygis* in the region. Recent studies also point to an emerging challenge caused by TB to elephants, as Asian elephants in Nepal have been shown to be infected with *M. tuberculosis* [22], elephants in India were seroreactive to TB antigens [23], and TB was detected from a wild elephant in Sri Lanka [24]. Additionally, we will discuss some case reports that describe TB in other wildlife. Thus, the second component of this review will be to address the conservation challenge caused by TB in wildlife of South Asia, with particular emphasis on *M. orygis*-associated TB and the emerging threat of TB in elephants.

http://dx.doi.org/10.5772/65798

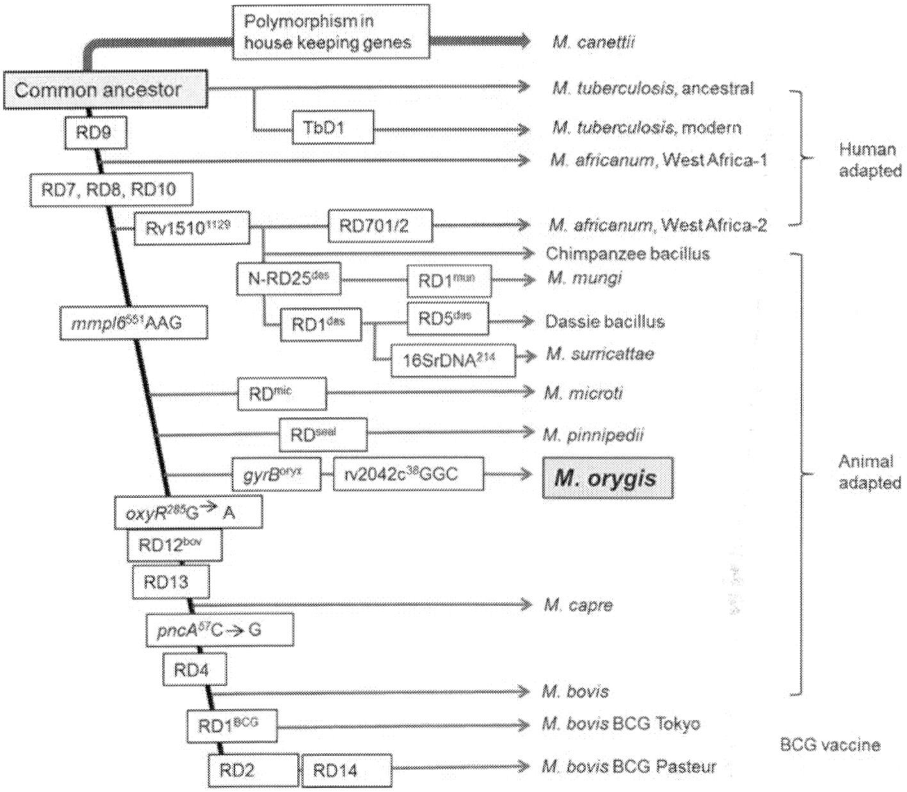

Figure 1. Updated phylogeny of *Mycobacterium tuberculosis* complex (MTBC). The phylogeny is based on the presence or absence of region of differences and single nucleotide polymorphisms and adapted from [10, 13, 39, 42, 65, 66].

This study reviews the current knowledge and information about animal-adapted members of the MTBC, the emerging threat of TB in wildlife, and the problem posed by TB to wildlife conservation in South Asia.

2. Animal-adapted members of MTBC

With the advent of a suite of molecular methods for genotyping of the MTBC, such as spoligotyping, deletion typing, and genome sequencing, in combination with increasing veterinary awareness for wildlife conservation, we now have the opportunity to scientifically pursue TB in wildlife. As a result, many new members of MTBC have been identified and the findings have clearly contributed to our understanding of the MTBC and their wildlife reservoirs (**Table 1**). Increasingly wildlife are recognized as reservoirs of these MTBC organisms and

MTBC species	Reservoir host	Geographical area	References
M. bovis	European badger	UK and Ireland	[67]
	Brush-tailed possum	New Zealand	[31]
	White-tailed deer	North America	[29, 30]
	Wild boar	Spain	[68]
	African buffalo	South Africa	[34]
	Lechwe	Zambia	[33]
M. caprae	Wild boar	Spain	[7]
M. pinnipedii	Seal, sea lion	Australia, Argentina	[8]
M. microti	Voles	UK	[9]
M. orygis	Antelopes*; deer, antelope§	Middle east*, South Asia§	[10, 16§, 20§]
M. suricattae	Meerkats	South Africa	[11, 39]
Dassie bacillus	Rock hyraxes	Southern Africa	[12, 36]
Chimpanzee bacillus	Chimpanzee†	Ivory Coast	[14]

*Antelope and oryx from Middle East countries were considered to be a source of transmission.
§Deer and antelope in South Asia may be a reservoir host.
†Only one case, may not be a reservoir host.

Table 1. *Mycobacterium tuberculosis* complex (MTBC) species and their wildlife reservoirs.

these findings pose a threat to wildlife conservation, undermine livestock TB control programs, and provide another avenue for zoonotic transmission.

M. bovis is the main cause of bovine TB but also infects many other domestic animals, wildlife, and also humans [15, 25]. It is globally distributed and one of the major causes of zoonosis in the geographic area where it is endemic. The WHO reported in 1998 that 3.1% of human TB cases were caused by *M. bovis* and 0–10% of sputum isolates from African patients could be from *M. bovis* [26]. A relatively recent study has crudely estimated 7 zoonotic TB cases/100,000 population/year in Africa [18].

M. bovis has successfully spread to several wildlife hosts and various geographical locations. The most well-known examples of wildlife involvement in the transmission of *M. bovis* infection are badgers in Great Britain which are involved in transmitting *M. bovis* to cattle [27], with a similar situation also evident in Ireland [28]. White-tail deer in Michigan and Minnesota in the United States are an important wildlife reservoir of *M. bovis* [29, 30], and the brush-tailed possum is New Zealand being another well-studied reservoir [31]. In all of the above examples, the wildlife reservoirs of bovine TB have greatly hindered the success of bovine TB eradication programs. While culling of these wildlife reservoirs has been carried out as a control option, such approaches are expensive and sometimes ineffective; indeed in one particular instance from the United Kingdom it was reported that badger culling increased TB in cattle [27]. Similarly, wild ungulates such as wild boar and red deer are maintenance hosts of *M. bovis* in the Iberian Peninsula, Europe and transmit *M. bovis* to other ungulates, carnivores, and livestock [32]. Lechwe, an antelope in Zambia, is another reservoir host of *M. bovis* that contributes to the spread of the infection to livestock [33]. Furthermore, Zambia has other flagship wildlife species such as elephants, lions, and hippos, to name a few, but the situation of TB in these species has yet to be ascertained. The strongest impact of *M. bovis* for wildlife conservation is probably in South Africa where the bacillus is maintained within African buffalo and has spilled over into 13 different wild animals including protected species like the lion, cheetah, and greater kudu [34]. Recently, TB caused by *M. bovis* in free-ranging rhinoceros in Southern Africa has been considered as an under recognized threat [35].

The MTBC species *M. caprae* was first isolated from domestic goats in Europe but has now been isolated from deer, wild boar, and livestock [7]. *M. pinnipedii* is a seal (pinnipeds)-specific member of the MTBC that has mostly been identified in animals from the Southern Hemisphere such as Australia and Argentina [8]. *M. microti* is a rodent MTBC mostly reported from voles in the UK and Europe [9]. There are three animal-adapted MTBC subspecies that are endemic to the Southern Africa subregion, namely, *M. mungi, M. suricattae,* and the Dassie bacillus [36]. *M. mungi* has been isolated from the banded mongoose from northern Botswana [13, 37], and its unique environmental mode of intra-mongoose transmission pathway has recently been elucidated [38]; however, this species has not yet been reported from other animals. The Dassie bacillus has been mostly isolated from rock hyraxes in Southern Africa [12]. *M. suricattae* is mostly reported from meerkats [11] and has a unique genetic feature in that it lacks the direct repeat locus and hence has no spoligotype pattern [39]. Interestingly, *M. mungi, M. suricattae,* and the Dassie bacillus have partial deletions of the RD1 locus, a locus that has been shown to be important for virulence of *M. tuberculosis* and *M. bovis* in various animal models (**Figure 1**).

Historically, it was believed that human TB evolved from bovine TB as a zoonosis. The hypothesis for that belief was based on the characteristic of a wide host range for *M. bovis* and a narrow host range limited to humans of *M. tuberculosis* [6]. However, from recent studies of whole genome and deletion analysis of *M. tuberculosis, M. bovis,* and other MTBCs, it has been shown that genome of *M. bovis* and other MTBCs including *M. orygis* has undergone numerous deletions relative to *M. tuberculosis* (**Figure 1**) indicating that *M. tuberculosis* predated the other MTBCs [6]. These successive losses of genetic material are postulated to have led to the appearance of different MTBCs that have become successful pathogens in certain hosts.

3. *M. orygis,* an emerging MTBC and a conservation threat in South Asia

M. orygis, also described as the oryx bacillus or the antelope clade, is a novel member and a subspecies of the MTBC with a unique phylogenetic position (**Figure 1**) that causes TB in animals and humans [10].

Before the use of molecular genotyping tools for differentiation of the members of the MTBC, *M. orygis* could have been misidentified as *M. tuberculosis* if isolated from infected humans, or *M. bovis* if isolated from infected animals. It was first reported in antelopes (oryx and waterbuck) at a zoo in the Netherlands, where the authors reported it originally as *M. bovis* but discussed its unusual feature of having a high copy number of the IS*6110* insertion sequence. In the same study, a human MTBC isolate having a similar IS*6110* pattern was identified, and both the human and animal isolates had an epidemiological link [40]. The clear molecular genetic distinction of *M. orygis* from other members of the MTBC was reported in 2005 [41]. While the subspecies name of this MTBC member was proposed as *orygis* to convey the fact that it was first characterized from oryx [10] or as an antelope clade to convey the group of animals from which it was most frequently isolated [6, 41–43], recently it has been isolated from many other animals and humans (**Table 2**). Thus, the isolation of *M. orygis* from a wide range of host species and the widespread geographical locations from where it has been isolated challenge previous concepts of it being an exclusively antelope clade and its host range and geographical distribution.

We reported the isolation of *M. orygis* from wild animals (a spotted deer and a blue bull) from a captive facility in Nepal [20]. We were aware of TB infections in the wild animals in this captive facility, and had assumed it to be caused by *M. bovis*; however, after molecular characterization, the TB isolates were confirmed to be *M. orygis*. Later in 2015, another *M. orygis* was isolated from a free-ranging greater one-horned rhinoceros in Chitwan National Park (CNP); the clinical history and necropsy of this animal suggested that it died from TB [16]. All three of these animals (spotted deer, blue bull, and rhinoceros) had extensive TB granulomatous lesions on the lungs, well encapsulated and filled with caseous necrotic material. The TB

Host (number of isolates)	Geographical location	Wild/captive/domestic	References
Antelope (2)	Netherlands	Captive	[10]
Water buck (3)			
Oryx (1)			
Antelope (1)	South Africa		
Deer (1)	United Kingdom		
Human (10)	South Asia	Not applicable	
Human (1)	South East Asia		
African buffalo (1)	South Africa	Captive	[69]
Cattle (1)	New Zealand	Domestic	[51]
Human (1)	New Zealand/Indian Immigrant	Not applicable	
Blue bull (1)	Nepal	Captive	[20]
Spotted deer (1)			
Rhinoceros (1)		Wild	[16]
Monkey (2)	Bangladesh	Captive/wild captured	[21]
Cattle (18)		Domestic	

Table 2. Reported cases of *Mycobacterium orygis* from different host species and geographical location.

lesions of the rhinoceros were limited to the lungs, whereas in the case of spotted deer and blue bull there were extrapulmonary lesions on lymph nodes, the gastrointestinal tract, and liver (**Figure 2**) [16, 20]. When cultured, all three *M. orygis* isolates yielded smooth and moist colonies unlike rough and dry colonies from *M. tuberculosis* (**Figure 3**). Although all three *M. orygis* isolates from these animals had the same spoligotyping pattern (SIT587), further molecular characterization by multilocus variable number of tandem repeat analysis (MLVA) revealed that both the deer and blue bull isolate had the same MLVA type while the rhinoceros isolate was a different type differing at one MLVA locus [16]. When we isolated *M. orygis* from the deer and blue bull from a captive wild animal facility in an earlier study [20], we had postulated that the origin of this TB might be from other animals in the CNP; the new finding of a slightly different strain of *M. orygis* from a free-ranging rhinoceros in CNP supports our original hypothesis [16].

The greater one-horned rhinoceros is the largest species of rhinoceros that is listed in Appendix I (most endangered) of the Convention on International Trade in Endangered Species (CITES), categorized as vulnerable by the International Union for Conservation of Nature (IUCN) Red List, and listed as a protected species by the Government of Nepal [44–46]. The present day free-ranging population of rhinoceroses in Nepal and India has increased from only 600 individuals in 1975 to 3555 individuals by mid-2015 [47]. In Nepal, the population of rhinoceroses is 645 individuals, of which 605 individuals live in CNP in a relatively narrow area of riverine grassland [48, 49]. From a conservation point of view, having a chronic and devastating disease like TB in this vulnerable and isolated population, that is already threatened from habitat destruction and poaching, is a matter of great concern for the animal's long-term survival. Also, CNP is listed by the United Nations Educational, Scientific and Cultural Organization (UNESCO) as a World Heritage Site because of its rich biodiversity, being home to globally protected animals such as Bengal tigers, Asian elephants, and greater one-horned

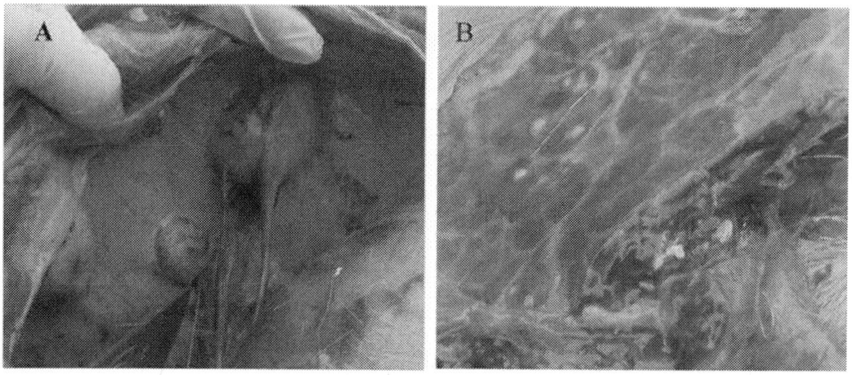

Figure 2. Description of tuberculosis lesions obtained during postmortem of a dead spotted deer. (A) Extrapulmonary. (B) Lung. Extrapulmonary tuberculosis lesions were of various sizes and capsulated with extensive liquefaction. Pulmonary tuberculosis lesions were of varying sizes from a single focal granuloma of 1–2 cm to extensive granulomatic lesion affecting a larger area of lung tissue (figure obtained with permission from Thapa et al. 2015 [20]).

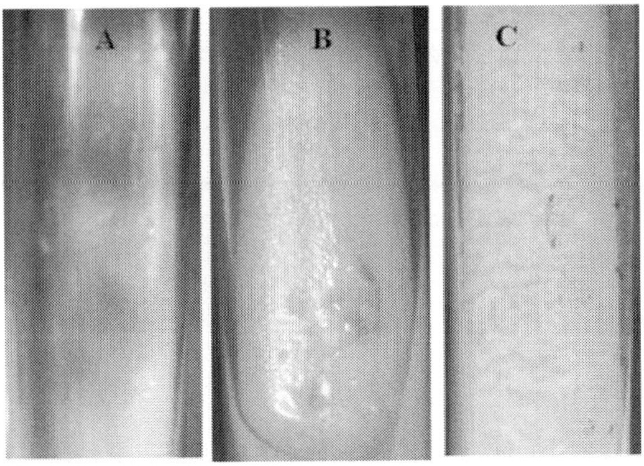

Figure 3. Comparative culture morphology of *M. orygis* and *M. tuberculosis* isolates. Culture of *M. orygis* isolated from the deer isolate (A), the blue bull isolate (B), and a comparative culture of a *M. tuberculosis* isolate from human (C). All the cultures are grown in Löwenstein-Jensen medium in the same laboratory but at different times (figure obtained with permission from Thapa et al. 2015 [20]).

rhinoceroses. The finding of TB in rhinoceroses in the park could also signify TB as a threat in other animals, including a strong possibility of unknown maintenance hosts of *M. orygis* in and around the national park [16]. CNP also shares an international border with the Valmiki National Park of India, and many CNP wildlife species, including rhinoceroses, routinely migrate between Nepal and India over this border. So, we cannot rule out the possibility of *M. orygis* infection being prevalent in wild animals in India.

In a study in India, 16 postmortem tissue samples from 25 suspected TB cases of spotted deer from a zoo were positive for IS*6110* PCR, indicating the presence of infection with members of the MTBC [50]. Since IS*6110* is present in all the MTBC species it cannot be used to differentiate the various members of the MTBC. Hence, as was the case in Nepal, the causative agent of TB in those deer from India could have been *M. orygis*. Therefore, *M. orygis*-associated TB in wild animals of South Asia may be a conservation threat and it should be further explored to ascertain its real impact on conservation of wild and protected animal species.

4. *M. orygis* may be endemically distributed in South Asia

In a recent study we isolated 20 *M. orygis* isolates from 18 cattle and two monkeys in Bangladesh [21]. All the cattle belonged to a farm and the two monkeys were from a zoo in Dhaka. TB lesions were identified in the lungs of the cattle and monkeys during postmortem examination. All of the cattle and monkey isolates shared the same spoligotyping pattern (SIT587) and the MLVA analysis divided the isolates into 3 clusters where the biggest cluster comprised 15 of the cattle isolates and the 2 monkey isolates. When compared with the larger cluster, one cattle isolate had a difference at one MLVA locus, whereas two cattle isolates had differences at three loci [21]. Despite not being able to trace the origin of the cattle on the farm, the finding of a large cluster with 15 cattle isolates and two monkey isolates indicated the widespread prevalence of this *M. orygis* strain in that area in both domestic and wild animals. Furthermore, the presence of other strains with one or three MLVA loci difference to the main cluster suggested a heterogeneous population of strains in that area. These finding suggest that *M. orygis* in that region of Bangladesh may be maintained in the cattle population, although we cannot rule out the possibility of an unknown reservoir host that would ultimately indicate the endemic presence of *M. orygis* in that area of Bangladesh.

From our studies in Nepal and Bangladesh, we have identified five MLVA types, two types in Nepal and three types in Bangladesh, in different wild and domestic animals. In Nepal, we speculated that there may be an unknown reservoir host of *M. orygis*, whereas in Bangladesh we suggest the possibility that cattle were a reservoir host. The genetic variation of *M. orygis* strains in different animals and geographical locations supports the view of its endemic distribution in South Asia. Adding further support to this latter hypothesis, 10 *M. orygis* isolates from South Asian human patients have been reported [10]. Similarly, *M. orygis* was also isolated from an Indian immigrant in New Zealand, with the probable origin of infection suggested to be in the patient's native country where there was opportunity of contact with local animals [51].

To the best of our knowledge, there are no confirmed diagnoses of *M. bovis* from human and cattle in Bangladesh although there are several studies of bovine TB in Bangladesh based on serological diagnosis [52]. Some of the cases of bovine TB could be from *M. orygis*. Similarly, wildlife TB in the South Asian region is largely understudied and probably underreported; even if TB lesions are observed during postmortem examination, confirmatory testing to identify species is generally not performed. Hence, these recent reports of *M. orygis* from South Asia present an important research question as to the origin and distribution of animal TB in South Asia; there could be a possibility that *M. orygis* may have predated *M. bovis* as a cause of TB in animals in South Asia, if the latter was introduced and rapidly expanded along with importation of European cattle. This assumption is supported by the discovery of different strains of *M. orygis* in wild animals in Nepal, the wide distribution of *M. orygis* strains in Bangladesh, and the potential links of *M. orygis* isolates to India. Thus, these findings support the One Health significance of *M. orygis* through the human-animal-environment connection as well as its endemic distribution in South Asia.

5. Elephant TB in South Asia

For centuries, elephants have been revered in Asia where they are the part of the region's culture and religion. They are seen as the guardians of forest, play a critical role to maintain the forest ecosystem, and are lately an indispensable asset for conservation work in Asia. Unfortunately, because of shrinking protected habitats and extensive poaching their numbers have declined and elephants are now classified as an endangered species [53]. As in other South Asian countries, elephants have sociocultural and economic value in Nepal [54]. However, elephants in Asia now face a potentially grave threat from TB.

TB in elephants is a reemerging disease caused primarily by *M. tuberculosis*, a human form of TB; however, infection with *M. bovis* has also been infrequently reported. Intensive study on elephant TB only received attention after 1996 when two circus elephants died of TB in the United States. Recently, TB in elephants is increasingly being detected in their host range countries and in zoological collections around the world [55].

TB screening in Asian elephants (*Elephas maximus*) has been carried out in South Asian countries including Nepal, India, and Sri Lanka, countries that also have a high TB prevalence in the human population. The tradition of keeping and training elephants in South Asia is long standing and continues to support various religious and tourism-related activities, as well as forestry and conservation management. This practice provides ample opportunity for transmission of TB from humans to elephants or vice versa.

There are about 200 captive elephants in Nepal that are used by government authorities for patrolling of protected areas, by the private sector for eco-tourism, and in some cases for wildlife research projects [56]. TB was first reported in captive elephants in Nepal in CNP in 2002 [57] and from 2002 to 2014, more than 10 elephants died of TB in Nepal. A comprehensive study was conducted in Nepal for screening of TB in 115 captive elephants; the results show that 15 of 115 (13%) elephants were reactive on the Elephant TB Stat Pak® assay [58]. The Elephant TB Stat Pak® is a licensed serological test developed by ChemBio Diagnostics, Inc., Medford, NY, USA, that uses a cocktail of several selected *M. tuberculosis* and/or *M. bovis* antigens (ESAT-6, CFP-10, and MPB83) to detect TB antibodies in elephants [59]. However, the antigens used are common for the MTBCs so they cannot differentiate MTBC subspecies including *M. orygis*. Also, the *M. tuberculosis* strain isolated from three of the cases in Nepal was identified as the same strain found in Nepalese human TB patients [22].

In 2007, a Nepal Elephant Healthcare and TB Surveillance Program was initiated to conduct regular screening of captive elephants. Similarly, in 2011, the Nepal Elephant Tuberculosis Control and Management Action Plan (2011–2015) was endorsed by the Government of Nepal providing detailed guidelines for the management of TB including the diagnosis and treatment of TB in elephants of Nepal [60]. This was the first elephant TB control action plan implemented in an Asian elephant range country.

A study conducted in India has shown that 15% of over 300 captive elephants tested were reactive on the Elephant TB Stat-Pak® assay. The highest seroreactivity was among temple elephants; these elephants had the greatest contact with humans of the three management groups studied [23]. *M. tuberculosis* was also isolated from two wild Asian elephants in India [61]. Similarly, *M. tuberculosis* was recently isolated from a wild elephant for the first time in Sri Lanka; however, the source of infection from possible human contact or potential wildlife reservoirs could not be traced for this elephant [24]. This finding is very important as it reflects the potential of TB transmission within wildlife populations in Sri Lanka.

Although, *M. tuberculosis* is usually identified from elephants, we cannot rule out the possibility of *M. bovis* or even *M. orygis* infection in elephants as all three MTBCs are endemic in South

Asia. TB is increasingly being identified in captive elephants in South Asian range countries and this directly poses a great risk of TB transmission from infected elephants to wild bull elephants that frequently visit captive herds for mating, to other endangered mammalian species, as well as to the people working with them. The regular screening of elephants for TB will help in prevention of this disease in elephants and susceptible hosts which will undoubtedly help in the conservation of this endangered species.

6. Other cases of TB in wildlife in South Asia

There are few reports of wildlife TB from South Asian countries. One study from a zoological collection in India reported TB in spotted deer [50], while a similar study in Pakistan reported TB in spotted deer and two antelopes, namely, chinkara gazella and black buck [62]. A few studies have reported TB from langur in India and rhesus monkey in Nepal [63, 64], and these may be associated with *M. tuberculosis* transmission from contact with humans as in South Asia as there are many feral monkeys residing in communities. However, TB in wildlife in South Asia has not been adequately studied to fully understand the impact. We hope that in the future an increased awareness of diseases in wildlife and better access to advanced diagnostic tests and technologies that allow for precise MTBC species identification will refocus efforts on the study of TB in wildlife and help to minimize or eliminate risks to species of conservation importance.

7. Conclusion

In summary, in this review we have provided an overview of animal-adapted members of the MTBC and indicated the role of wildlife as a reservoir host. In the future, the study of wildlife TB will not only be important for conservation efforts but also for finding potential novel subspecies of the MTBC. We also discussed in detail reports of *M. orygis*-associated TB in wildlife of South Asia, highlighting recent studies from Nepal and Bangladesh. The finding of *M. orygis* in a free-ranging rhinoceros, an endangered species, with evidence for the wide distribution of *M. orygis* attests to its threat for wildlife conservation in the region. The evidence of cattle as a potential reservoir host of *M. orygis* in South Asia warrants further analysis and increased surveillance in livestock so as to mitigate the risk to public health. Our review of elephant TB from different countries demonstrates another threat to the conservation of indigenous wildlife in the region. The cultural and socioeconomic values of elephants, in addition to their conservation importance, support the need for continued surveillance and appropriate management and response to this disease threat. Given the paucity of research in this area, with a low number of publications, we think it evident that wildlife TB is overlooked and underexplored. The topics discussed here and the paucity of scientific studies signify a greater underlying problem. We sincerely recommend that further studies be pursued as a matter of urgency to explore the threat of TB to wildlife in South Asia.

Author details

Jeewan Thapa[1], Chie Nakajima[1,2,*], Kamal P. Gairhe[3], Bhagwan Maharjan[4], Sarad Paudel[5], Yogendra Shah[1], Susan K. Mikota[6], Gretchen E. Kaufman[7], Deborah McCauley[7], Toshio Tsubota[5], Stephen V. Gordon[8] and Yasuhiko Suzuki[1,2,*]

* Address all correspondence to: cnakajim@czc.hokudai.ac.jp and suzuki@czc.hokudai.ac.jp

1 Division of Bioresources, Hokkaido University Research Center for Zoonosis Control, Sapporo, Hokkaido, Japan

2 The Global Station for Zoonosis Control, Hokkaido University Global Institution for Collaborative Research and Education, Sapporo, Japan

3 Chitwan National Park, Chitwan, Nepal

4 Nepal Anti-Tuberculosis Association, German-Nepal Tuberculosis Project, Kathmandu, Nepal

5 Graduate School of Veterinary Medicine, Hokkaido University, Sapporo, Japan

6 Elephant Care International, Hohenwald, Tennessee, USA

7 Veterinary Initiative for Endangered Wildlife, Bozeman, Montana, USA

8 UCD School of Veterinary Medicine, University College Dublin, Dublin, Ireland

References

[1] Olson DM, Dinerstein E. The Global 200: Priority ecoregions for global conservation. Annals of the Missouri Botanical Garden. 2002;89(2):199–224.

[2] Myers N, Mittermeler RA, Mittermeler CG, da Fonseca GAB, Kent J. Biodiversity hotspots for conservation priorities. Nature. 2000;403:853–858.

[3] Morese SS, Mazet JAK, Woolhouse M, Parrish CR, Carroll D, Karesh WB, Zambrana-Torrelio C, Lipkin I, Daszak P. Prediction and prevention of the next pandemic zoonosis. The Lancet. 2012;380(9857):1956–1965.

[4] Gordon SV, Behr MA. Comparative Mycobacteriology of the *Mycobacterium tuberculosis* complex. In: Mukundan H, Chambers M, Waters R, Larsen M, editors. Many Hosts of Mycobacteria: Tuberculosis, Leprosy, and other Mycobacterial Diseases of Man and Animals. Oxfordshire, UK: CAB International, 2015.

[5] de Jong BC, Antonio M, Gagneux S. *Mycobacterium africanum*—review of an important cause of human tuberculosis in West Africa. PLoS Neglected Tropical Diseases. 2010;4:e744. http://dx.doi.org/10.1371/journal.pntd.0000744

[6] Brosch R, Gordon SV, Marmiesse M, Brodin P, Buchrieser C, Eiglmeier K, Garnier T, Gutierrez C, Hewinson G, Kremer K, Parsons LM, Pym AS, Samper S, van Soolingen

D, Cole ST. A new evolutionary scenario for the *Mycobacterium tuberculosis* complex. Proceeding of the National Academy of Science of the United States of America. 2002;19:3684–3689.

[7] Rodriguez S, Bezos J, Romero B, de Juan L, Alvarez J, Castellanos E, Moya N, Lozano F, Javed MT, Saez-Llorente JL, Liebana E, Mateos A, Dominguez L, Aranaz A. *Mycobacterium caprae* infection in livestock and wildlife, Spain. Emerging Infectious Disease. 2011;17:532–535.

[8] Cousins DV, Bastida R, Cataldi A, Quse V, Redrobe S, Dow S, Duignan P, Murray A, Dupont C, Ahmed N, Collins DM, Butler WR, Dawson D, Rodriguez D, Loureiro J, Romano MI, Alito A, Zumarraga M, Bernardelli A. Tuberculosis in seals caused by a novel member of the *Mycobacterium tuberculosis* complex: *Mycobacterium pinnipedii* sp. nov. International Journal of Systematic and Evolutionary Microbiology. 2003;53:1305–1314.

[9] Cavanagh R, Begon M, Bennett M, Ergon T, Graham IM, de Hass PEW, Hart CA, Koedam M, Kremer K, Lambin X, Roholl P, van Soolingen D. *Mycobacterium microti* infection (Vole tuberculosis in wild rodent populations. Journal of Clinical Microbiology. 2002;40:3281–3285.

[10] van Ingen J, Rahim Z, Mulder A, Boeree MJ, Simeone R, Brosch R, van Soolingen D. Characterization of *Mycobacterium orygis* as *M. tuberculosis* complex subspecies. Emerging Infectious Disease. 2012;18:653–655.

[11] Parsons SDC, Drewe JA, Nicolaas C, van Pittius G, Warren RM, van Helden PD. Novel cause of tuberculosis in Meerkats, South Africa. Emerging Infectious Disease. 2013;19:2004–2006.

[12] Cousins DV, Peet RL, Gaynor WT, Williams SN, Gow BL. Tuberculosis in imported hyrax (*Procaviacapensis*) caused by an unusual variant belonging to the *Mycobacterium tuberculosis* complex. Veterinary Microbiology. 1994;42:135–145.

[13] Alexander KA, Laver PN, Michel AL, Williams M, van Helden PD, Warren RM, Gey van Pittius, NC. Novel *Mycobacterium tuberculosis* complex pathogen, *M. mungi*. Emerging Infectious Disease. 2010;16:1296–1299.

[14] Coscolla M, Lewin A, Metzger S, Maetz-Rennsing K, Calvignac-Spencer S, Nitsche A, Dabrowski PW, Radonic A, Niemann S, Parkhill J, Couacy-Hymann E, Feldman J, Comas I, Boesch C, Gagneux S, Leendertz FH. Novel *Mycobacterium tuberculosis* complex isolate from a wild chimpanzee. Emerging Infectious Disease. 2013;19:969–976.

[15] de Lisle GW, Bengis RG, Schmitt SM,O'Brien DJ. Tuberculosis in free-ranging wildlife: Detection, diagnosis and management. Revue Scientifiqueet Technique (International Office of Epizootics). 2002;21(2):317–334.

[16] Thapa J, Paudel S, Sadaula A, Shah Y, Maharjan B, Kaufman GE, McCauley D, Gairhe KP, Tsubota T, Suzuki Y, Nakajima C. *Mycobacterium orygis*-associated tuberculosis in free-ranging rhinoceros, Nepal, 2015. Emerging Infectious Disease. 2016;22(3):570–572.

[17] STAC. SAARC epidemiological response on tuberculosis. SAARC Tuberculosis and HIV/AIDS Center, Kathmandu, Nepal, 2014.

[18] Muller B, Salome D, Alonso S, Hattendorf J, Laisse CJM, Parsons SDC, van Helden PD, Zinsstag J. Zoonotic *Mycobacterium bovis* induced tuberculosis in humans. Emerging Infectious Disease. 2013;19:899–908.

[19] Prasad HK, Singhal A, Mishra A, Shah NP, Katoch VM, Thakral SS, Singh DV, Chumber S, Bal S, Aggarwal S, Padma MV, Kumar S, Singh MK, Acharya SK. Bovine tuberculosis in India: Potential basis for zoonosis. Tuberculosis. 2005;85:421–428.

[20] Thapa J, Nakajima C, Maharjan B, Poudel A, Suzuki Y. Molecular characterization of *Mycobacterium orygis* isolates from wild animals of Nepal. Japanese Journal of Veterinary Research. 2015;63(3):151–158.

[21] Rahim Z, Thapa J, Fukushima Y, Suzuki H, van der Zanden, Adri GM, Gordon SV, Suzuki Y, Nakajima C. Tuberculosis caused by Mycobacterium orygis in dairy cattle and captured monkeys in Bangladesh: a new scenario of tuberculosis in South Asia. Transboundary and Emerging Diseases. 2016. DOI: 10.1111/tbed.12596.

[22] Paudel S, Mikota SK, Nakajima C, Gairhe KP, Maharjan B, Thapa J, Poudel A, Shimozuru M, Suzuki Y, Tsubota T. Molecular characterization of *Mycobacterium tuberculosis* isolates from elephants of Nepal. Tuberculosis. 2014;94:287–292. doi: 10.1016/j.tube.2013.12.008

[23] Abraham D, Cheeran JV, Sukumar R, Mikota SK, Rao S, Ganguly S, Varma S. Health assessment of captive Asian elephants in India with special reference to tuberculosis. Report to Project Elephant, Ministry of Environment and Forests, Government of India, New Delhi, 2008.

[24] Perera BVP, Salgadu MA, Gunawardena GSPS, Smith NH, Jinadasa HRN. First confirmed case of fatal tuberculosis in a Wild Sri Lankan elephant. Gajah. 2014;41:28–31.

[25] Fitzgerald SD, Kaneene JB. Wildlife reservoirs of bovine tuberculosis worldwide: Hosts, pathology, surveillance and control. Veterinary Pathology. 2012;50:488–499.

[26] Cosivi O, Grange JM, Daborn CJ. Zoonotic tuberculosis due to *Mycobacterium bovis* in developing countries. Emerging Infectious Disease. 1998;4:59–70.

[27] Donnelly CA, Woodroffe R, Cox DR, Bourne J, Gettinby G, Le Fevre AM, McInerney JP, Morrison I. Impact of localized badger culling on tuberculosis incidence in British cattle. Nature. 2003;426:834–837.

[28] Olea-Popelka F, Flynn O, Costello E, McGrath GE, Collins JD, O'Keeffe JJ, Kelton DF, Berke O, Martin SW. Spatial relationship between *Mycobacterium bovis* strains in cattle and badgers in four areas in Ireland. Preventive Veterinary Medicine. 2005;71:57–70.

[29] Carstensen M, DonCarlos MW. Preventing the establishment of a wildlife disease reservoir: A case study of bovine tuberculosis in Minnesota, USA. Veterinary Medicine International. 2011;2011:Article ID 413240, 10 p.

[30] O'Brien DJ, Schmitt SM, Fitzgerald SD, Dale EB. Management of bovine tuberculosis in Michigan wildlife: Current status and near term prospects. Veterinary Microbiology. 2011;151:179–187.

[31] Caley P, Hone J, Cowan PE. The relationship between prevalence of *Mycobacterium bovis* infection in feral ferrets and possum abundance. New Zealand Veterinary Journal. 2001;49:195–200.

[32] Santos N, Correia-Neves M, Almeida V, Gortázar C. Wildlife Tuberculosis: A Systematic Review of the Epidemiology in Iberian Peninsula. In: Cunha M, editors, Epidemiology Insights. ISBN: 978-953-51-0565-7. Rijeka, Crotia: InTech, DOI: 10.5772/33781. 2012. Available from: http://www.intechopen.com/books/epidemiology-insights/wildlife-tuberculosis-a-systematic-review-of-theepidemiology-in-the-iberian-peninsula [Accessed: 2016-08-04].

[33] Hang'ombe MB, Munyeme M, Nakajima C, Fukushima Y, Suzuki H, Matandiko W, Ishii A, Mweene AS, Suzuki Y. *Mycobacterium bovis* infection at the interface between domestic and wild animals in Zambia. BMC Veterinary Research. 2012;8:221.

[34] Michel AL, Bengis RG, Keet DF, Hofmeyr M, de Klerk LM, Cross PC, Jolles AE, Copper D, Whyte IJ, Buss P, Godfroid J. Wildlife tuberculosis in South African conservation areas: Implications and challenges. Veterinary Microbiology. 2006;112:91–100.

[35] Miller M, Michel A, van Helden P, Buss P. Tuberculosis in rhinoceros: An under recognized threat? Transboundary and Emerging Diseases. 2016. DOI: 10.1111/tbed.12489

[36] Clarke C, van Helden P, Miller M, Parsons S. Animal-adapted members of the *Mycobacterium tuberculosis* complex endemic to the southern Africa subregion. Journal of the South African Veterinary Association. 2016;87(1):a1322. http://dx.doi.org/10.4102/jsava.v87i1.1322

[37] Alexander KA, Pleydell E, Williams MC, Lane EP, Nyange JF, Michel, AL. *Mycobacterium tuberculosis*: An emerging disease of free-ranging wildlife. Emerging Infectious Disease. 2002;8:598–601.

[38] Alexander KA, Sanderson CE, Larsen MH, Robbe-Austerman S, Williams MC, Palmer MV. Emerging tuberculosis pathogen hijacks social communication behavior in the group-living banded mongoose (*Mungos mungo*). mBio. 2016;7(3):e00281–16. DOI:10.1128/mBio.00281-16

[39] Dippenaar A, Parsons SDC, Sampson SL, van der Merwe RG, Drewe JA, Abdallah AM, Siame KK, Gey van Pittius NC, van Helden PD, Pain A, Warren RM. Whole genome sequence analysis of *Mycobacterium suricattae*. Tuberculosis. 2015;95:682–688.

[40] van Soolingen D, de Hass PE, Haagsma J, Eger T, Hermans PWM, Ritacco V, Alito A, van Embden JDA. Use of various genetic markers in differentiation of *Mycobacterium bovis* strains from animals and humans and for studying epidemiology of bovine tuberculosis. Journal of Clinical Microbiology. 1994;32:2425–2433.

[41] Mostowy S, Inwald J, Gordon S, Martin C, Warren R, Kremer K, Cousins D, Behr MA. Revisiting the evolution of *Mycobacterium bovis*. Journal of Bacteriology. 2005;187:6386–6395.

[42] Huard RC, Fabre M, de Hass P, Lazzarini LCO, van Soolingen D, Cousins D, Ho JL. Novel genetic polymorphisms that further delineate the phylogeny of the *Mycobacterium tuberculosis* complex. Journal of Bacteriology. 2006;188:4271–4287.

[43] Smith NH, Kremer K, Inwald J, Dale J, Driscoll JR, Gordon SV, van Soolingen D, Hewinson RG, Smith JM. Ecotypes of the *Mycobacterium tuberculosis* complex. Journal of Theoretical Biology. 2006;239:220–225.

[44] Convention on International Trade in Endangered Species. Appendices [Internet]. 2015. Available from: https://www.cites.org/eng/app/appendices.php [Accessed: 2015-11-22].

[45] Talukdar BK, Emslie R, Bist SS, Choudhury A, Ellis S, Bonal BS, Malakar MC, Talukdar BN, Barula M. *Rhinoceros unicornis*. The IUCN Red List of Threatened Species 2008. 2008. e.T19496A8928657. Available from: http://dx.doi.org/10.2305/IUCN.UK.2008.RLTS.T19496A8928657.en [Accessed: 2016-08-04].

[46] Department of National Park and Wildlife Conservation. Protected species (mammals) [Internet]. 2015. Available from: http://www.dnpwc.gov.np/protected_species/species/mammals [Accessed: 2015-10-30].

[47] World Wildlife Fund. Greater one horned rhino. 2015. Available from: http://http://wwf.panda.org/what_we_do/endangered_species/rhinoceros/asian_rhinos/indian_rhinoceros/[Accessed: 2015-10-30].

[48] NTNC. National rhino count 2015: 645 individual rhinos in Nepal [Internet]. 2015. Available from: http://www.ntnc.org.np/news/national-rhino-count-2015-645-individual-rhinos-nepal [Accessed: 2015-10-30].

[49] Subedi N, Jnawali SR, Dhakal M, Pradhan NMB, Lamichhane BR, Malla S, Amin R, Jhala YV. Population status, structure and distribution of the greater one-horned rhinoceros *Rhinocerosunicornis* in Nepal. Oryx. 2013;47:352–360.

[50] Sharma S, Mallick GP, Verma R, Ray SK. Polymerase chain reaction (PCR) amplification of IS6110 sequences to detect *Mycobacterium tuberculosis* complex from formalin-fixed paraffin-embedded tissue of deer (*Axis axis*). Veterinary Research Communication. 2007;31:17–21.

[51] Dawson KL, Bell A, Kawakami RP, Coley K, Yates G, Collins DM. Transmission of *Mycobacteiumorygis* (*M. tuberculosis* complex species) from a tuberculosis patient to a dairy cow in New Zealand. Journal of Clinical Microbiology. 2012;50:3136–3138.

[52] Mahmud MAA, Belal SMSH, Shoshe NZ. Prevalence of bovine tuberculosis in cattle in the selected Upazila of Sirajganj district in Bangladesh. Bangladesh Journal of Veterinary Medicine. 2014;12:141–145.

[53] World Wildlife Fund. Asian elephants. 2016. Available from: http://wwf.panda.org/what_we_do/endangered_species/elephants/asian_elephants/ [Accessed: 2016-07-18].

[54] Thapa J. Twin elephants born in Nepal. Gajah. 2009;30:53.

[55] Mikota, SK, Lyashchenko KP, Lowenstine L, Agnew D, Maslow JN. Mycobacterial infections in elephants. In: Mukundan H, Chambers, MA, Waters WR, Larsen MH, editors, Many Hosts of Mycobacteria. Tuberculosis, Leprosy, and Other Mycobacterial Diseases of Man and Animals. CABI Publishing House, Nosworthy Way, Wallingford, UK, 2015. pp. 259–276.

[56] Pradhan NMB, Williams AC, Dhakal M. Current status of Asian elephants in Nepal. Gajah. 2011;35:87–92.

[57] Gairhe K. A case study of tuberculosis in captive elephants in Nepal. Report submitted to Department of National Parks and Wildlife Conservation, Kathmandu, Nepal, 2002.

[58] Mikota SK, Gairhe K, Giri K, Hamilton K, Miller M, Paudel S, Lyashchenko K, Larsen RS, Payeur JB, Waters WR, Greenwald, Dumonceaux G, Vincent B, Kaufman GE. Tuberculosis surveillance of elephants (*Elephas maximus*) in Nepal at the captive-wild interface. European Journal of Wildlife Research. 2015;61:221–229.

[59] Lyashchenko KP, Greenwald R, Esfandiari J, Olsen JH, Ball R, Dumonceaux G, Dunker F, Buckley C, Richard M, Murray S, Payeur JB, Andersen P, Pollock JM, Mikota S, Miller M, Sofranko, Waters WR. Tuberculosis in elephants: Antibody responses to defined antigens of *Mycobacterium tuberculosis*, potential for early diagnosis, and monitoring of treatment. Clinical and Vaccine Immunology. 2006;13:722–732.

[60] Nepal Elephant Tuberculosis Control and Management Action Plan (2011–2015). Government of Nepal, Ministry of Forests and Soil Conservation, Department of National Parks and Wildlife Conservation, Kathmandu, Nepal, 2011.

[61] Zachariah A. Emerging diseases in the single largest Asian elephant (*Elephas maximus indicus*) population, Nilgiri Biosphere Reserve, South India. Report to United States Fish and Wildlife Service Asian Elephant Conservation Fund,2012.

[62] Shahid AL, Javed MT, Khan MN, Cagiola M. Prevalence of bovine tuberculosis in zoo animals in Pakistan. Iranian Journal of Veterinary Research, Shiraz University. 2012;13:58–63.

[63] Parmar SM, Jani RG, Kapadiya FM, Sutariya DR. Status of tuberculosis in the free living hanuman langur (*Presbytis entellus*) of Gujarat state. Indian Veterinary Journal. 2013;90:74–75.

[64] Wilbur AK, Engel G, Rompis A, Putra IGAA, Lee BPYH, Aggimarangsee N, Chalise M, Shaw E, Oh G, Schillaci MA, Jones-Engel L. From the mouths of monkeys: Detection of *Mycobacterium tuberculosis* complex DNA from buccal swabs of synanthropic Macaques. American Journal of Primatology. 2012;74:676–686.

[65] Gordon SV, Bottai D, Simeone R, Stinear T, Brosch R. Pathogenicity in the tubercle bacillus: Molecular and evolutionary determinants. Bioessays. 2009;31:378–388.

[66] Bos KI, Harkins KM, Herbig A, Coscolla M, Weber N, Comas I, Forrest SA, Bryant JM, Harris SR, Schuenemann VJ, Campbell TJ, Majander JK, Wilbur AK, Guichon RA,

Steadman DLW, Cook DC, Niemann S, Behr MA, Zummarraga M, Bastida R, Huson D, Nieselt K, Young D, Parkhill J, Buikstra JE, Gagneux S, Stone AC, Krause J. Pre-Columbian mycobacterial genomes reveal seals as a source of New World human tuberculosis. Nature. 2014;514:494–497.

[67] Gallagher J, Clifton-Hadley RS. Tuberculosis in badger: A review of the disease and its significance for other animals. Research in Veterinary Science. 2000;69:203–217.

[68] Gracia-Jimenez WL, Benuitez-Medina JM, Fernandez-Llario P, Abecia JA, Gracia-Sanchez A, Martinez R, Risco D, Ortiz-Pelaez A, Salguero FJ, Smith NH, Gomez L, Hermoso De Mendoza J. Comparative pathology of the natural infection by *Mycobacterium bovis* and *Mycobacterium caprae* in wild boar (*Susscrofa*). Transboundry and Emerging Disease. 2013;60:102–109.

[69] Gey van Pittius NC, Perrett KD, Michel AL, Keet DF, Hlokwe T, Streicher EM, Warren RM, van Helden PD. Infection of African buffalo (*Synceruscaffer*) by oryx bacillus, a rare member of the antelope clade of the *Mycobacterium tuberculosis* complex. Journal of Wildlife Disease. 2012;48:849–857.